above
Main Loan Desk, University of Texas, Austin.

above right
Poets and printers combine talents at the 1830 Acorn-Smith hand press for *Lime Kiln Press* productions, U.C. Santa Cruz Library.

right
Main Library, Stanford University.

below right
A William Blake collection is part of U.C. San Diego's Mandeville Department of Special Collections.

below
Reference Room, Main Library, U.C. Berkeley.

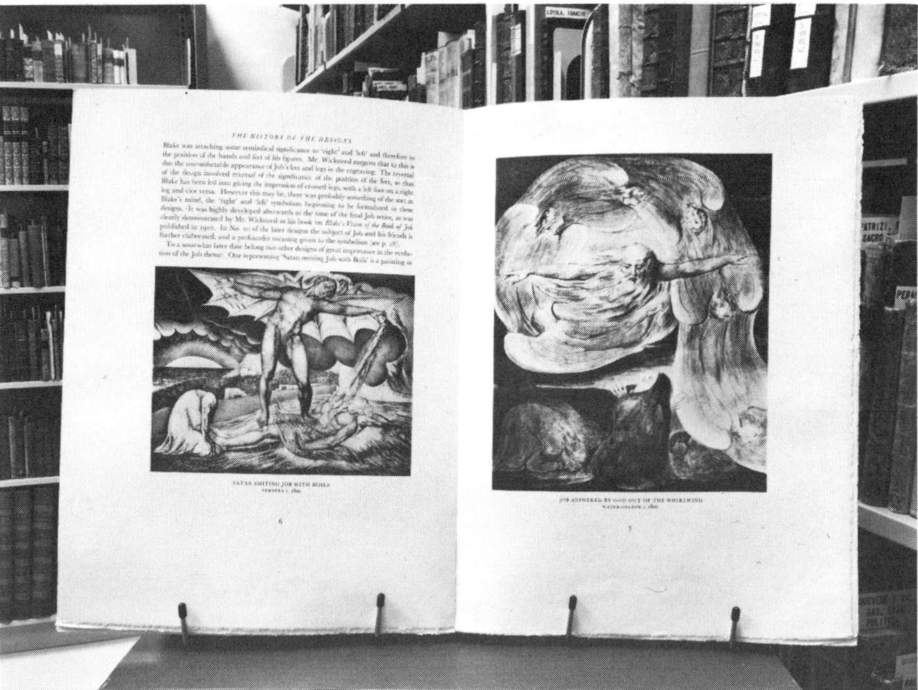

THE HISTORY OF THE DESIGNS

SATAN SMITING JOB WITH BOILS

JOB ANSWERED BY GOD OUT OF THE WHIRLWIND

Managing the Library Fire Risk

Baker Library, Harvard
Business School

Clark Library, University of
California at Los Angeles

Meyer Undergraduate Library,
Stanford University

Managing the Library Fire Risk

By
JOHN MORRIS

UNIVERSITY OF CALIFORNIA
Second Edition, 1979

Morris, John, 1912–
Managing the library fire risk.

Bibliography: p.
1. Library buildings—Fires and fire prevention.
I. Title.
TH9445.L5M67 1979 628.9'22 78-22603
ISBN 0-9602278-1-4

LIBRARY OF CONGRESS CARD CATALOG NUMBER 78-22603

SECOND EDITION
PUBLISHED BY THE
OFFICE OF RISK MANAGEMENT AND SAFETY
UNIVERSITY OF CALIFORNIA
BERKELEY, CA 94720
$14.00
PRINTED IN THE UNITED STATES OF AMERICA BY
THE UNIVERSITY OF CALIFORNIA PRINTING DEPARTMENT

Preface to the Second Edition

As long as libraries have been in existence, they have been faced with the threat of fire, and many distinguished libraries have been wasted by destructive fires. A few years ago the University of California undertook a program of improving fire protection of important libraries, and in doing so sought to provide librarians with authentic information on library fires and methods of protection against fire. The standard manual then available was not current, in that since its publication library fire disasters had occurred, innovations had been made in salvage of wet books, and new fire defense hardware and systems had been developed. To report these events and bring the record of library fire experience up to date *Managing the Library Fire Risk* was published in late 1975.

The second edition has been considerably expanded with new information covering the years 1975–1978, including the chronicle of library fire and water casualties. Fire defense systems of great new libraries are described, as well as the improvement of older libraries. The personal safety of patrons and staff is given more adequate treatment than in the earlier edition.

Illustrations showing proprietary hardware and systems for fire protection are used only because they fairly represent the type of protection shown without making extravagant claims; most of these things are available from a number of manufacturers or suppliers. The use of these illustrations is not to be considered as a recommendation or an endorsement of any system or product by the author or by the University.

Managing the Library Fire Risk, Second Edition, 1979 is a project of the Office of Risk Management and Safety, University of California.

STEPHEN J. DENESS
Systemwide Risk Manager
University of California
November, 1978

John Morris
November, 1978

ACKNOWLEDGEMENTS

Our thanks for information and good advice are due a number of friends in the fields of library management, architecture, fire protection and publishing, all of whom contributed to the writing of one or another edition of *Managing the Library Fire Risk:* William H. Adamson, Chester Babcock, Roland Bellman, Robert Beth, Anita Blair, Donald J. Boehmer, Philip R. Brook, Ed Carey, Robert Carington, James F. Casey, Nina Caspari, Margaret Chartrand, Donald H. Cloudsley, Ellen Cornish, Barbara Coyle, Cecelia Duncan, Robert Eustachy, Anson Finley, David J. Fischer, Don Fustich, Herbert F. Gladney, J. T. Glass, Bruce K. Harvey, Thora Hutchison, Rudolph A. Imarata, Carolyn E. Jakeman, Harlan Kessel, Richard J. King, Lewis G. Kirk, David Kosakowski, Raymond Kozar, Harry C. Lein, Stephen Lesnak, Peter Lockhart, Lois Lundquist, James McLaird, Dany Mestre, Keyes D. Metcalf, Jean C. Morris, Herbert F. Mutschler, Kenneth Nesheim, Irvin D. Nicholas, Russell M. Norman, Karl Nyren, Gyo Obata, W. H. Overend, Richard Peacock, Frazer Poole, George Reese, Jr., Bruce Reynolds, Elizabeth Reynolds, Noel Savage, Loren Sapp, John C. Shively, William Slemmer, John R. Smith, Kenneth Smith, Isabel Sorrier, Erwin Surrency, Stanley Tarr, Paul Teague, Kenneth Turton, Evans Walker, Lamar Wallis, Rex Wilson. Our special thanks to Feona J. Hamilton of The Library Association, London, for information on library fires in England, and to Dick Darrell and the *Toronto Star* for the picture of a fire at the University of Toronto.

DISCLAIMER

The Regents of the University of California and the author deny formally any liability for the use or misuse of information published in this book, and for errors and omissions which might originate from using it as a guide in construction or in any other way. It is not meant to be used as an engineering manual, nor as a substitute for the professional counsel in fire protection engineering which is available almost everywhere to librarians, library trustees and architects.

Contents

1.

2.

3.

4.

5.

Public Libraries

1. San Francisco
2. Galesburg, Illinois
3. Enoch Pratt Free Library, Baltimore
4. Berkeley, California
5. Parkside Branch, San Francisco

I.

Fire in the Library

"A model library building will discharge three offices—it will store the books in absolute safety from fire in an accessible manner, with light, air and an average temperature, and with provision for indefinite expansion . . . Nearly every library building ever constructed fails in all these particulars . . . It is scarcely an exaggeration to say that no single library building now standing discharges all three . . ."

Talcott Williams
"Plans for the Library Building of the University of Pennsylvania 1888.[1]

Sedgewick Library, University of British Columbia, Vancouver
Opened January 1973; completely protected with automatic sprinklers

FIRE IN THE LIBRARY

As long as libraries have existed they have been prey to destruction by fire. Twenty-five great libraries were burned during the last 2,500 years by armies of one country invading another, not to mention those destroyed by air bombardment of 1938–1945 in England and Europe.[2] Many others have been burned in fires of accidental origin or set by arsonists.

New England was still a colony of the British Crown when the library of Harvard College was struck by fire. Harvard Hall burned to the ground in January 1764, with most of its 5,000 books. John Harvard's personal collection was entirely lost, with the sole exception of *"Christian Warfare Against the Deuill, World and Flesh."*[3] Since that time fires have damaged or destroyed libraries from year to year, but there has never been an incidence of fires so frequent as to distinguish libraries as a bad risk for insurance, or one in which there is a fire hazard to life of any alarming degree.

During the night of August 24–25, 1814, the troops of British General Robert Ross burned the Library of Congress, then very small, in the course of setting fire to the principal capital buildings in Washington. Librarian Peter Magruder resigned under fire when a congressional committee claimed that he could have avoided the loss by obtaining wagons to remove the books before the assault.[4] The private library of Thomas Jefferson was later purchased to replace the destroyed books. The library burned again in 1825, with "heavy damage to the collection." It burned once more in 1851 as a consequence of "unattended fireplaces" with a loss of 35,000 volumes.[5]

Duplicate Records

The Hingham, Massachusetts library burned in 1878, and payment of insurance on the books was delayed because all records which could substantiate the loss were also burned in the fire. Shortly afterwards, LIBRARY JOURNAL advised its readers to avoid this predicament "because library fires are now much in vogue."[6] This is still good advice. Any unique records or set of files can be duplicated and stored in a separate place, removing a possibility of devastating loss from fire or vandalism.

The Free Library of Birmingham, in Warwickshire, Shakespeare's county, burned in January, 1879. Lost in the fire were a Shakespeare collection of 8,000 volumes, a very fine Cervantes collection and the Staunton library of manuscripts, books, pamphlets, prints and antiquities, all related to the history of Warwickshire and collected by three generations of the Staunton family. Destroyed were 336 English editions of Shakespeare and 91 in other languages, along with "posters, placards, pamphlets, caricatures"[7] and many other things. The fire was directly attributed to negligence of a workman using an open flame to thaw out a gas pipe during alterations to the building; unfortunately the managers of the library had not realized the importance of taking special precautions to protect it from the hazards of the job. Fire hazards of contractor operations have been the origin of a number of library fires, but none more destructive than the Birmingham incident.

Melvil Dewey, dean of American library professionals, developed a keen appreciation of the fire danger. Whereas in 1891 he described the quarters occupied in the capitol buildings at Albany by the New York State Library as "absolutely fireproof,"[8] his awareness of the fallacy of this description grew year by year. In his report to the State Legislature asking for better fire protection for the library he said:

Dewey Recognized the Danger

> "The Capitol walls are so massive that we have no fear of fire except as it might burn out individual rooms finished in wood. Hundreds of thousands of feet of oak have been used in shelving and interior finish, and in spite of careful installation of electric wires, we cannot avoid the fear that some day this woodwork in some room will be accidentally set on fire and priceless material destroyed. The scientific explanation of how the fire occurred may be perfect, but the fact that rats or mice gnawed off insulation or that workmen accidentally broke it with their saws (as has happened a score of times in the past dozen years) might tell how it happened, but will not replace our lost treasures . . . In our manuscript room are collections which have cost the state vast sums and which money could not replace, and yet there is no place to keep them except a room honeycombed with oak and interlaced with electric wires."

For lack of fire protection improvements recommended by Dewey and others the library burned in a spectacular fire in November, 1911. "Flying sheets of scorched paper—historical manuscripts and pages out of books—flecked the sky over Albany like snowflakes and drifted to earth over a 20-mile radius." Totally destroyed were 450,000 books, 270,000 manuscripts, 300,000 pamphlets and a general card catalog of nearly one million cards which had been compiled over a 20-year span. A 78-year-old watchman died in the fire.

This fire was called "the greatest catastrophe of modern library annals," since it reduced to ashes a large part of what was then the fifth largest library in the nation and among the first 20 in the world.

The public library in St. Paul, Minnesota burned April 27, 1915 with a loss of $450,000. The fire started in the basement of a store in the library building. Among the materials lost were two manuscripts of Professor W. D. Johnston, Vol. 2 of his "Life History of the Library of Congress," and an important professional work on university libraries, which he had to abandon as a result of the fire.[9]

Libraries in universities have been struck by fire many times. Omitting earthquake-caused fires and burnings related to conquest, there have been at least

49 college and university fires of significance, a number of which resulted in total destruction. Some of these are of special interest because of the circumstances of the fire or the nature of the loss.

Fire leveled a 3-story stone, wood-joisted college building in Winfield, Kansas April 16, 1950. Two students found the blaze shortly after midnight in a janitor storage room in the basement. They fought it with hand extinguishers before deciding to call the fire department. Housed in the building were chemistry and theater libraries and an art collection along with other college facilities. The loss was $298,000.

The Wheaton College Library fire in December 1960 at Norton, Massachusetts was started by a contractor's cutting torch in a newspaper room. The loss was $52,000.

Two spectacular fires destroyed university libraries situated in towers. On August 10, 1965 at Austin, Texas, workmen were using an acetylene torch at the 20th floor level, and sparks ignited papers stacked in a storage room below. Damage from smoke, water and debris was extensive throughout the 27-story tower. A relatively small area was burned out, mainly industrial designs and projects, framed portraits and those books closest to the point of origin of the fire. The University of Texas fine collections of theater art and literature survived the fire, except for some of the Harry Houdini and P.T. Barnum items in the Kendall collection. Many books required a careful salvage processing for weeks afterward to remove smoke and water, and the cleanup and repair job for the tall building was a formidable and costly operation.

Only a few months later, on April 19, 1966 in New York City, an incendiary fire caused great destruction in the tower library of the Jewish Theological Seminary of America.*

Dakota Wesleyan's library at Mitchell, S.D. was completely destroyed and two students killed when College Hall burned in 1888. The building that replaced it burned in 1966 and again the library was destroyed; the loss included irreplaceable Indian relics and a highly prized taped interview.

A very similar experience was that of St. Francis College in Loretto, Pennsylvania, six miles from any fire department, whose library was completely destroyed by fire in 1888 and again in 1958, with the loss of very old and cherished books. Destroyed in the 1958 fire, along with a 45,000-volume college library, were personal libraries, a master's thesis, 300 slides prepared for a doctor's thesis and two rare books printed before 1500.[10]

Two New York City library fires which were suppressed by automatic protection are reported by the National Fire Protection Association.[11] In January, 1965 fire broke out in the stacks in a 10-story university library of fire-resistive construction and completely sprinklered. "One sprinkler operated and put out the fire before firemen arrived in response to an alarm transmitted from the sprinkler system. Loss: $2,500." In February, 1951 fire that started in the

* This fire and other set fires in libraries are described in Section III, ARSON.

stockroom of a college library was put out by a single sprinkler head with a loss of $250.

Defective wiring in a heater control was believed to be the point of origin of a fire at the Lewis College Library, Lockport, Illinois, January 15, 1959. The loss was $200,000. The sole protection for the library was a watchman who patrolled every three or four hours.

We know that 19th century "fireproof" construction and various superior constructions in use today have relatively little value for protecting valuable contents of buildings unless automatic early-warning alarms or extinguishing systems are also present. The fire protection profession began to develop in 1874 with the introduction of the first successful automatic systems of extinguishment and early detection of fire. Great advances made in these systems, in fire protection theory and hardware over the period 1874–1975 have made it possible to build a new library today or equip an existing library with an armament of fire defenses that reduces the threat of a severe fire loss almost to zero.

REFERENCES

1. Talcott Williams, "Plans for the Library Building of the University of Pennsylvania," LIBRARY JOURNAL, vol. 13, no. 8, August 1888:238.
2. "Protecting the Library and its Resources," American Library Association, Chicago, 1963–64:216–230.
3. Rene Kuhn Bryant, "Harvard University Library 1638–1968," Harvard University Library, Cambridge, Mass., 1969:11.
4. Henry Steele Commager, "Jefferson and the Book Burners," American Heritage, vol. IX, no. 5, August 1958:65–68.
5. "Protecting the Library and its Resources":220.
6. LIBRARY JOURNAL, vol. 4, no. 2, February 28, 1879:63.
7. LIBRARY JOURNAL, vol. 4, no. 1, January 1879:19–20.
8. Cecil R. Roseberry, "A History of the New York State Public Library," State Education Department, Albany, 1970:81.
9. LIBRARY JOURNAL, vol. 40, no. 6, June 1915:385–6.
10. "Occupancy Fire Record—Libraries," National Fire Protection Association, Boston, Mass., 1961:7.
11. Recommended Practice for the Protection of Library Collections from Fire, NFPA 910, 1970:910–14.

II.

The Library Fire Risk Today

"I object, in the fifth place, to this plan of construction, on account of its insecurity from fire. In an interior finished with wood, no arrangement could be more skilfully devised for favoring the destructive operations of fire than a series of alcoves piled one upon the other six stories high, with every facility for draft, unless it be a pile of empty packing cases. When a building of this kind takes fire the work of the insurance adjuster is very simple, for it is a total loss of the whole library . . ."

William Frederick Poole
"The Construction of Library Buildings," 1881.

Central Library, University of California, San Diego

THE LIBRARY FIRE RISK TODAY

An indispensable source of library fire risk information is the 1963 Library Technical Project, "Protecting the Library and its Resources."[1] What appears to warrant a new publication is a series of developments over the years 1964–1978 affecting library fire protection:

1. Several great libraries have been wasted by destructive fires.
2. Values of old books, the cost of new books and the cost of library construction have increased sharply, and it is likely that these values will continue to rise.
3. The threat of arson has become far more serious; the 1973 total of 94,300 set fires reported for the U.S. was nearly twice the 1968 figure. Several libraries have reported arson losses and attempts.
4. The fire protection industry has developed new systems, concepts and hardware, making automatic extinguishment more efficient and reducing the threat of water damage to books.
5. The science of the salvage of books and other materials wet from fire-fighting operations and natural disasters has been greatly advanced.

Two Schools of Thought

There have been two schools of thought about fire protection for libraries. One group[2] reasons that a book is hard to ignite, burns slowly, and left alone will stop burning. They do not readily endorse the suggestion that funds be diverted into reconstruction for greater fire safety; they are reluctant to install automatic sprinklers because of concern for water damage to books through some sort of malfunction of a system, or even incidental to putting out a fire.

The other group sees the typical older library as essentially a warehouse full of costly, ready-to-burn materials. They see open stairways and ventilating slots as flues which during a fire would conduct smoke and superheated fire gases upward through the stacks and create a holocaust. This group is generally supported by fire protection engineers, insurance underwriters, and fire marshals.

Library Fire Tests—An Historical Review

Experiments have been carried out periodically to prove one or another point about books and fire. Fire and water tests using obsolete law books were made outdoors in England as early as 1884[3] to determine what containers would give "books of exceptional value" the best protection in fire, and it was decided that boxes with numerous relatively small breather holes would give good fire protection and yet not admit too much water from fire fighting operations; boxes with larger vent holes did not provide protection as effective.

Tests of library materials in fire were made by the Bureau of Standards in

1925[4] in connection with studies of fireproofing. Fires were set with a loading of 85 lbs. per square foot of books and papers under carefully controlled conditions. The researchers were astonished at the intensity of the fire they got, which reached a temperature just short of 2,500 degrees Fahrenheit. Whereas most of their tests were over in two hours, this fire lasted through the following afternoon, when the debris was still glowing hot. The test building was damaged in that the roof was warped upward 2½ inches and the masonry walls perceptibly bulged out of line.

Factory Mutuals fire laboratories at Norwood, Massachusetts conducted carefully documented library book stack burns in 1959–60[5] which appeared to prove (a) that books in book stacks will develop a raging fire, and (b) automatic sprinklers will control such a fire with only moderate wetting to books not actually being burned but close to the fire. The test was reported in fire protection and library professional journals. In the annual report of the New York Public Library 1959–60[6] the tests are described as sponsored by the library to see if automatic sprinkler protection might cause "more irreparable damage than fire." The conclusion: "There appears to be no question about the fact that *under conditions such as those which exist in the Central Building,* where book stacks have been installed in a multi-tiered construction rather than on fire-resistant floors, water would do less overall damage than fire . . ."

George Schaefer of Cornell University observed the Norwood fire tests.[7] ". . . Within a matter of five minutes the fire was at the third level and was a blazing inferno . . . In nine minutes the entire four levels were ablaze and the temperature at the top was 1400 degrees. The metal shelving bent and melted with the heat and the uprights began to sag. It was a sickening sight as you realized what could happen and how wrong so many people were when they stated books would not burn and water was worse than fire . . . Forever is a long time. It seemed very short as I watched the conflagration on December 17 and realized that what had taken many years and centuries to develop might be gone within 30 minutes."

Fire Resistive Construction

Another fire test of importance was made at Cornell University in 1963,[8] when it was decided to determine through applied research whether the new Olin Library could safely be built without installing automatic sprinklers. The test seemed to indicate that with metal shelving, and with no combustibles for fuel other than the books themselves, the fire risk was moderate in a *tightly compartmented stacks construction of the most modern fire resistive design.* Their fires only flickered by comparison with the roaring flames obtained in the open construction tests at Norwood. Their conclusions:

1. Stacks installed on fire resistant structural floors are to be preferred to multi-tiered stack construction.
2. Sheet steel shelves are to be preferred to either wood or U-bar steel shelves.

3. Metal trucks are to be preferred to wood trucks on stack floors.
4. All openings between floors (e.g., stairways, book lifts and elevators) should be enclosed with fire resistant materials.
5. An approved, properly installed fire detection system connected to a fire department should be installed.

The ALA Technical Project writers had some reservations[9] about the Norwood tests, based on the failure of the laboratories' engineers to establish identical physical conditions for the sprinklered and unsprinklered stack setups, and on other points. The Cornell test was sometimes referred to as having disproved the Norwood tests; this is not correct, of course, since the constructions involved were quite different. In any event, if there were still any uncertainties about the vulnerability of open stacks to fire or the capability of sprinklers to control library fires, these were to be resolved in actual fire disasters and incidents during the years following.

The term "fire resistive construction" can be misleading. A building that is all reinforced concrete, protected steel, brick and stone will not by virtue of these materials alone guarantee life safety nor superior protection of property from fire. As long ago as 1879 the fallacy of "fireproof" construction was pointed out in a paper by Cornelius Walford presented to a conference of librarians in England.[10] He reported that the fire of 1878 that nearly totally destroyed the Birmingham Free Library proved that an "absolutely fireproof structure" was no guarantee against fire disaster. "It is only assurance," he said, "that the building itself will survive." Other factors were seen as more essential than the "fireproof shell" in library construction in an editorial[11] on building plans for the Library of Congress:

> "Emphasis is laid upon the fireproof character of the walls, floors and shelving; upon the power of isolating each division of the library in case the books—the only combustible things—get on fire . . ."

Likewise, when William Frederick Poole described an ideal library plan to the American Library Association in 1879[12] he listed seven main points. Number one was "The building shall be constructed in compartments, and as nearly fireproof as is possible, so that if fire starts, it shall be confined in the compartment in which it originates, and the rest of the library be saved."

Compartmentation is still an important theory of fire protection; it does not prevent destruction by fire, but tends to limit it to a single compartment; where the theory succeeds, it prevents small fires from becoming big fires. Poole likened this to the division of a ship's hull into watertight sections; in a fire small openings in the walls, or a fire door not tightly shut, will leak heat and flame into another area just as openings in the ship's bulkheads would permit water to leak through. To the extent that one is willing to accept the risk of loss of the contents of any single unit of space in a fire resistive structure, compartmentation alone is an acceptable device. It is far better as an accessory to automatic protection systems.

The vulnerability of "fireproof" and "fire resistive" constructions to fire has been shown in numerous disasters, such as the Morgan Annex Post Office fire of December 15, 1967, a $10,000,000 loss, as well as many other costly fires, some involving large loss of life. It has become evident through these disasters that construction with reinforced concrete and protected steel in buildings *without additional features for life safety and the control of fire* provides little protection either for persons or contents.

What should a library have to pay for adequate fire protection? How much fire protection is enough? These questions have to be countered with another, "How important is your library?" If it is the central resource of a university or college, or has special collections, or irreplaceable items of great value, or materials with a unique value of some kind to that particular library, this is a library demanding a superior system of fire protection.

The ALA Technical Project handbook[13] provided criteria for judgment as to whether a sprinkler system or other improvements should or should not be installed in a library. These criteria are not less valid today. The library that has fire resistive construction refined with tight compartmentation, good fire department backup protection and a modern central station early warning system may be sufficiently protected; but libraries with multi-tiered open stacks, large areas, vertical openings between levels, and complicated exit patterns can hardly afford to be without the manifest benefits of a complete system of automatic extinguishment. The decision is an important one and deserves the attention of a fire protection engineer.

REFERENCES

1. "Protecting the Library and its Resources," American Library Association, Library Technical Project, Chicago, 1964.
2. Dorothea M. Singer, "The Insurance of Libraries," American Library Association, Chicago, 1946:3.
3. "Preservation of Books from Fire," LIBRARY JOURNAL, vol. 9, no. 11, November 1884:191.
4. Engineering News-Record, vol. 94, no. 19, May 7, 1925:763.
5. P. O. Cotton, "Library Book Stack Fire Tests." NFPA Quarterly, vol. 53, no. 4, April 1960:289–295.
6. "Annual Report of the New York Public Library, 1959–60" in Bulletin of the New York Public Library, vol. 64, no. 12, December 1960:618.
7. George L. Schaefer, "Fire!", LIBRARY JOURNAL, vol. 85, no. 3, February 1, 1960: 504–5.
8. H. B. Schell, "Cornell Starts a Fire," LIBRARY JOURNAL, vol. 8, no. 17, October 1, 1960:3398–9.
9. "Protecting the Library and its Resources," 1964:207–215.
10. LIBRARY JOURNAL, vol. iv, no. 11, November 1879:414–5.
11. LIBRARY JOURNAL, vol. 9, no. 11, November 1884:187.
12. Arthur E. Boswick, "The Library and Its Home," New York, 1933:31.
13. "Protecting the Library and Its Resources," 1964:94.

Arson

"Available statistics indicate that incendiary ignitions now account for seven percent of all fires, and ten percent of all fire losses, but the true number may be closer to three times those fractions . . ."

Kendall D. Moll
"Arson, Vandalism and Violence," 1974.

The April, 1966 fire disaster of the Jewish Theological Seminary, New York

ARSON

Arson in 1975 is more than ever before important as a potential source of destructive fire and a factor to be weighed in considering the fire risk of libraries. The National Fire Protection Association estimated there were 94,300 incendiary fires in 1973, nearly twice the 1968 total; the actual number is suspected to be even greater, since in many fires where the evidence of arson may have been destroyed the judgment of the investigator has had to be "cause unknown or undetermined." Likewise in Great Britain where the arson investigator saw an average case load of 580 arson incidents in the 1950's, in 1969 he saw almost 2,300. Here again the total is considered much greater than reported, since "arson is not expertly investigated in many jurisdictions."[1]

The arsonist is not easy to recognize. He may be a person with a disturbed mind, with a pathological grievance based on political or ideological grounds, or a person acting under the influence of drugs or stimulants. His motivation may be entirely obscure and the crime directed against the library only because it is more accessible and vulnerable than some other public building. Most libraries are open to anyone, and with many areas not closely supervised. Since the arsonist usually strikes when the library is closed, preventing him from breaking into the building at night or hiding out in the building after closing time is the first line of defense against arson.

Representative library fires of incendiary origin include the following:

The Bureau of Government library of the University of Michigan was set afire in the afternoon of June 6, 1950 by an instructor. It burned with a loss estimated at $637,000. Materials lost included 33,000 books and 17,000 "irreplaceable items." There was no automatic system either to suppress the fire or provide early warning, and it spread rapidly through the building and into the large attic. Firemen used 850,000 gallons of water to put out the flames.

The State Library and Office Building, Lansing, Michigan, burned February 8–13, 1951. This was essentially a fire resistive building. The fire was set by an employee in records stored on open shelves. Flames spread through an undivided 43,000 sq. ft. room, only slightly smaller than a football field. The fire burned for five days. The top two floors were so badly damaged they had to be removed. Loss: $2,850,000.

Public Library, Bayonne, New Jersey, May 31, 1959: A 24-year-old man entered the building at night by forcing a window; he ran through the stacks waving a burning corn broom, set several fires, then jumped out an upper window when trapped by smoke and heat. He then pulled a box fire alarm and hid. The firemen responded, but did not see the fire and went back to headquarters. A neighbor then telephoned the fire department; they came back and put out the fire, but not before the roof had collapsed. Thousands of books were either burned or thoroughly wet. Loss: $800,000.

Jewish Theological Seminary Library, New York, April 1966: A 12-story tower housed one of the world's most distinguished libraries of Jewish theology. It had brick walls with only a few small windows. Floors were reinforced concrete on steel beams. A small elevator and one open stairway provided access to individual floors.

Multi-tiered bookstacks extended continuously from the third to the twelfth story, and there were slot-like openings entirely around the stacks on each floor. This design permitted movement of the air for ventilation and heating, and when fire struck it let fire and smoke travel upward by the same route. There were neither automatic sprinklers nor automatic fire detection.

Fire, believed incendiary, started on the tenth story. There was a 20-minute delay in calling the fire department while the building maintenance crew tried to put out the fire. The fire department had great difficulty in getting hoses up to the fire and could find no way of venting the intense heat, smoke and steam.

'A Financial, Historical and Sentimental Loss'

Sixty firemen were overcome. They used nine hose streams for several hours to get the fire under control. Fire heavily damaged three upper floors, and water poured down through the lower stories. Approximately 70,000 books were burned and 150,000 soaked. "The fire resulted in a tremendous financial, historical, and sentimental loss of books, manuscripts, and records, many of them irreplaceable."[2] Loss: $3,000,000.

University libraries were more subject to vandalism and arson than others during the militant mob-scene years 1967–1970 in Europe and America alike; a number of great libraries suffered vandalism and fire atacks. Concerning "student" militant arson, Stephen Barlay reported[3] that in 1970 in France "Maoists, Communist splinter groups, ideologically non-committed vandals, students, political agitators posing as students, and right wing organizations soon claimed a share of the glory." The University of California at Berkeley had a fire bombing and a destructive fire March 9, 1970, in a library reading room actually occupied by students at the time. The Berkeley fire department responded promptly to find a raging fire in furnishings and combustible interior finish materials, and only by a desperate attack prevented the fire from reaching the very large areas of book stacks. The loss was heavy, but fortunately limited to furniture, equipment, walls and floors; there was no important damage to books.

The Ortega Branch Library in San Francisco was set afire and burned one night in April, 1972 with a loss of about $50,000. The fire department was able to confine their attack to books that were actually on fire and avoid wetting others. A neighborhood youth gang was tentatively blamed for the fire.

During the night of July 2, 1972 an arsonist broke into the Long Beach, California main library and set a fire which was not discovered until morning; firemen put down the fire skillfully and without wetting undamaged books, but more than 5000 items were lost to fire, including a cherished California history collection and some old and rare publications. The damage to building and contents was more than $150,000. Exactly a month later police investigating a break-in at the library arrested an "unemployed artist and dancer," age 26, who admitted setting the library fire and other fires in obedience to the dictates of his religion, which required sacrificial offerings as a black magic ritual. He was convicted of the library fire and sentenced to prison for from 2 to 20 years.

Book Drops Need Protection

The Fremont Branch of the Seattle Public Library System was damaged by fire during the night on January 14, 1974, when someone broke in through the front door and set nine or more fires; the loss amounted to $4000. A far greater loss in the same area was the nearly complete destruction of the Federal Way Branch in the King County Library System March 29, 1975.[4] Two boys started the fire by throwing a burning match book into the book drop late at night. The cost of restoration was more than $365,000 for the building alone.

Book collections were largely destroyed, except for 14,000 volumes in circulation at the time of the fire. Although book drops of all 40 libraries in the system were promptly strengthened against arson, two adolescent boys were able to start another fire in the Carnation Branch on a Sunday in broad daylight; fortunately it was quickly discovered and put out with very little damage to the library.

Centerville Library in Fremont, California was destroyed June 1, 1974 when a 24-year-old man under the influence of drugs broke in and set fire to furnishings. Firemen entered a side door and attacked the fire from inside the building so as to minimize water damage to books. The loss to the building and contents was estimated at $20,000.

Arson in the seventies continues to be an important and increasing threat. Kendall Moll of Stanford Research Institute, making a study[5] of arson in the United States, found that reported incendiary fires increased fourfold from 1950 to 1960, and 13 times by 1970. He identifies eight types of arson:

1. Fraud—usually to collect insurance.
2. Political fires, including bank burnings and "protests."
3. Pyro fires (pyromaniacs)—The root causes are emotional and sexual problems of youth, destructive and sadistic impulses.
4. Crime "coverup," burglary, auto theft; destructive "trashing" or vandalism motivated by race hatred.
5. Spite—as sometimes occurs after verbal exchanges between youth and firemen at a fire.

6. Vanity—the fire setter becomes a hero, putting out fire or rescuing victims from a fire which he himself has set.
7. Psycho—the arsonist is on alcohol, drugs or amphetamines and wants to watch the pattern of the flames.
8. Vandalism—destruction of vacant buildings either as a protest or as "instant urban renewal."

Arson is a "wild card" factor which must be figured into fire protection calculations along with experience with so-called "accidental" fire sources, such as electrical faults. The set fire does not always smolder for hours, providing opportunity for an automatic detection system to identify it and send a signal; it is more likely to flash into a full-blown fire at the moment it is touched off. The possibility of arson places a premium on automatic extinguishment for managing the library fire risk.

University of Illinois, Champaign recorded an incident in early 1970 in which a fire bomb set off at night in a classroom building was extinguished by a single sprinkler head and the fire department summoned by the automatic waterflow alarm. At Macy's main store in New York City in October, 1969 an arsonist set five incendiary devices in various parts of the store. Four of them detonated and started fires; all were promptly extinguished by automatic sprinklers.[6]

REFERENCES

1. Stephen Barlay, "Fire, an International Report," Brattleboro, Vermont, 1969: 213.
2. "Recommended Practice for Protection of Library Collections from Fire," National Fire Codes, vol. 15, NFPA 910, 1975:435.
3. Barlay, "Fire, An International Report," 1969:212.
4. "Arson in King County," LIBRARY JOURNAL, vol. 100, no. 10, May 15, 1975.
5. Kendall Moll, "Arson, Vandalism and Violence: Law Enforcement Problems Affecting Fire Departments," U.S. Government Printing Office, Stock #27000–00251, March 1974.
6. Barlay, "Fire, An International Report," 1969:213.

IV.

The Philadelphia Story

"Some people tell you books do not burn, but they do. Also, when they get wet they expand, and they pop out exactly like popcorn."

Erwin C. Surrency
"Guarding Against Disaster," 1973.

The fire that struck this great university library had many of the elements of the ultimate library disaster. It is a valuable point of reference in estimating the maximum fire risk of any library, and particularly one of open construction and not protected by automatic fire systems. This article is reprinted in full from the November, 1972 FIRE JOURNAL, copyrighted by the National Fire Protection Association. Reprinted by permission.

The Charles Klein Law Library Fire

Philadelphia, Pennsylvania

A. ELWOOD WILLEY, *NFPA Fire Record Department*

On July 25, 1972, fire destroyed the Charles Klein Law Library at Temple University in Philadelphia, Pennsylvania. The fire, which originated in a rear office area, spread through large-area concealed spaces under the roof. The library collection suffered heavy damage, and the structure was destroyed, with collapse of the roof into the main reading room. With assistance from Library of Congress preservation experts, a significant amount of the 400,000-volume collection was salvaged. Delayed detection, large-area combustible concealed spaces, and lack of automatic sprinkler protection were factors in this loss.

The author is indebted to the Philadelphia Fire Department for assistance during his visit to the city to collect data on this loss. The assistance of Temple University officials is also greatly appreciated. Peter Waters, Restoration Officer, Preservation Office, Library of Congress, provided valuable salvage information.

Unless otherwise credited, the photos are by the Philadelphia Fire Department.

The photo below shows the front of the building after roof collapse.

The Charles Klein Law Library was located on the campus of Temple University, at 1700 North Broad Street, Philadelphia. Originally built in 1891 as a synagogue, the structure was acquired by the University in 1953. In 1959 it was renovated for use as a library. The renovated three-story-and-basement building was divided into three principal areas. Offices were located on the first and second floors, with a cockloft above at the front (west side) of the building (see the diagram at right). A street-floor library reading room

was located in the center of the building. There was a large-area concealed space located above the reading-room ceiling on the second- and third-floor levels. Law School offices were located on all three floors at the rear (east side) of the building.

The 400,000-volume library collection was housed in the basement and the main reading room. Library offices, a cataloging department, and a book repair section were located on the first floor in the front two-story section. There were other University offices on the second floor of the section.

The library collection included reference volumes of state and Federal laws, legal periodicals, and rare documents on English and American law. Rare books and documents, including papers of Benjamin Franklin, were located in the basement. The 21,000-volume Lucas Hirst Collection was located on the first floor in a separate southwest corner room.

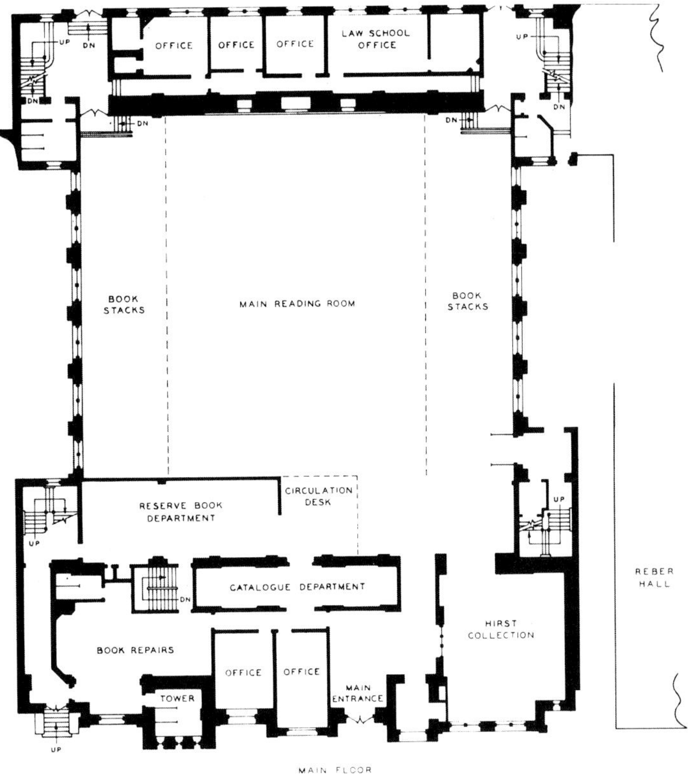

Right: floor plan

Below: over-all map, showing position of apparatus and hose streams

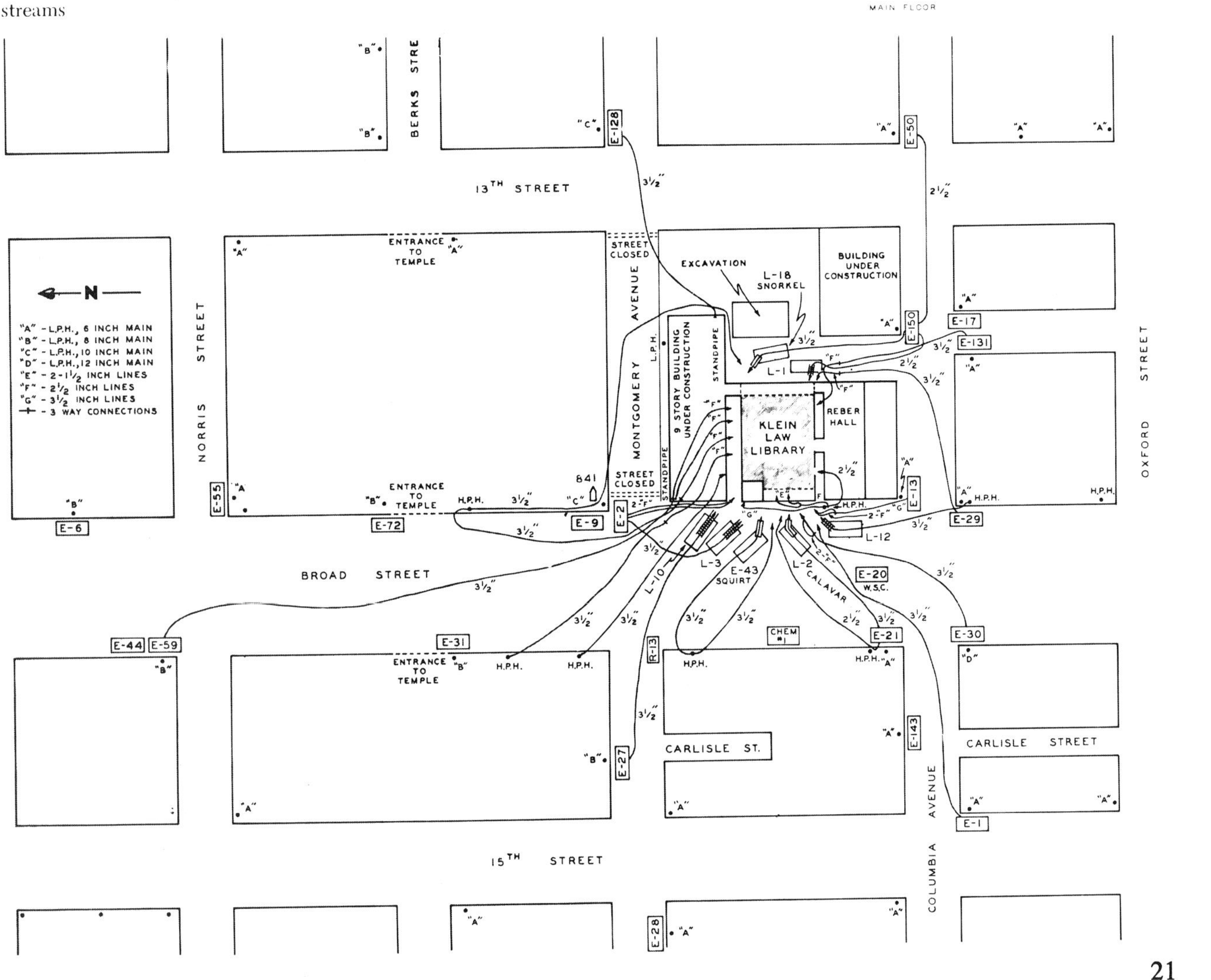

The Klein Law Library was situated adjacent to a new law library under construction to the north. Reber Hall, housing additional School of Law offices, adjoined the library to the south. Direct access to Reber Hall was provided through basement and first-floor corridors and a second-floor open-air crossover. The basement and first-floor openings to Reber Hall were protected with single sliding fire doors for a Class A location installed in the masonry walls. A brick and concrete-block wall separated the library offices from the main reading room. The openings in this wall were protected with rolling steel fire doors for a Class A location. A solid-core wood door was also installed in a wall opening between the cataloging room and the reading room.

A wood-lath and plaster-on-wood-stud wall separated the reading room and the three-story Law School office section. Wood-panel doors with glass provided access from the reading room to the first-floor office corridor. Steel doors were installed in second-floor openings in this wall. These doors provided access to a former balcony area above a more recently installed suspended ceiling. A 120-foot tower dominated the façade of the building in the northwest corner.

The 152-by-170-foot structure was of ordinary construction, with brick bearing walls with a stone veneer. The floors were wood on wood joists. The hip-type roof consisted of wood plank with a metal covering supported with purlins on heavy timber truss. The interior partitions were plaster on wood lath. The interior wall and ceiling finishes were plaster with some wood wainscot paneling in office corridors.

The changes during building renovation included a new pan-type reinforced concrete floor slab for the main reading room. The slab was four inches thick, with 12-inch joists. A new plaster-and-wire-lath ceiling was installed above the reading room. The ceiling was installed in a suspension system on ¼-inch metal rods. The floor-to-ceiling height in the first-floor stack areas was eight feet. The ceiling height above the reading room was 16 feet. The large-area concealed space under that portion of the roof formerly over the synagogue sanctuary was subdivided with two one-hour-rated partitions, which reportedly ran to the underside of the roof deck.[1] The previously described rolling steel fire doors

[1] Construction details were not available.

Above: the ruins of the Klein Law Library, looking toward the southwest. Undamaged Reber Hall is in the background. Some of the debris had been removed from third-floor offices and the reading room before this photo was taken.
Right: basement book stacks. Note mold damage in the two bottom shelves.
Far right: book stacks in the main reading room.
NFPA FIRE RECORD DEPT.

were installed to subdivide the first-floor areas. The fire partitions and fire doors had been required by the city building officials for renovation plans approval.

The building was served by five stairways. Two stairs in the rear (east side) ran from the basement to the third floor. One stair in the front (west side) ran from the basement to the first floor. Two other stairs on the west side ran between the street and the second floor. The stairs were partially enclosed with solid-core doors installed at various locations. The east-side stairways discharged to a first-floor corridor that led to the outside. The stairways on the west side also discharged to a first-floor corridor. An exit from this corridor was located at the main library entrance at the front. Egress from the main reading room was provided by a main west-side exit and by two doors opening to the office corridor at the rear. Emergency lighting and exit signs were installed in the building.

Books were stored on metal racks seven feet high. Each storage compartment was three feet long, with an average of five compartments per tier. Each compartment contained seven shelves spaced one foot apart vertically. The aisles between tiers varied from 32 to 40 inches.

A manual local-only fire alarm system was installed in the building. It was arranged to sound an evacuation alarm. There were portable fire extinguishers located throughout the building. The building was not sprinkler-protected.

During the afternoon of Tuesday, July 25, approximately 30 people and 11 staff members were in the library. The Law School offices at the rear were not occupied. Shortly before 2 pm a construction worker outside at the rear of the building saw smoke coming from the roof. The Philadelphia Fire Department received a telephone alarm at 1:45 pm. A full box alarm assignment – four engine companies and two ladder companies – was dispatched. Campus security forces were also alerted, and they responded to investigate. Occupants of the library were unaware of the fire until security officers activated the evacuation alarm. The security officers proceeded to the third floor and located the fire in a wall at the northeast corner of the building. Extinguishers were obtained, but the security officers could not get them to operate.

Led up the northeast stairs to the fire in the third-floor wall, first-arriving fire fighters attacked and knocked down visible fire on the third floor. However, heavy smoke continued to envelop the area. Ladder company personnel opened up walls and ceilings and also ventilated the roof above the area, to locate the source of the fire. At 2:05 pm a second alarm was sounded. Ladder companies on the roof found heavy fire showing above the library ceiling, and at 2:30 a third-alarm assignment was dispatched. The second-alarm assignment summoned four engine companies and one ladder company; the third alarm, four engine companies and two ladder companies.

As more smoke billowed from the roof of the building, University staff members and students began to realize the magnitude of the fire situation. An impromptu brigade of faculty, students, and onlookers was formed to salvage books from the first floor, making a human chain from the front of the main reading room out to the street as books were passed outside. In this manner the public card catalog and approximately 3,000 volumes were saved. The books were placed in an auto agency building across the street from the library. Fire officers did not permit salvage volunteers to enter that portion of the main reading room under the ceiling, which was suspended 16 feet above the floor, because of the collapse hazards. As the roof weakened, parts of roof members could be heard dropping to the ceiling, and the chief officer soon ordered everyone out of the building. At approximately 2:35 the ceiling and roof collapsed and at 2:37 the fourth alarm was sounded. The fourth-alarm assignment brought an additional four engine companies and another ladder company. Master streams from ladder pipes and elevated platform apparatus were placed in service along with hand lines to contain the fire within the now heavily involved structure. At the height of the fire, it is estimated, the fire flow was 11,000 gpm.

The enervating effects of the hot and humid weather conditions[2] necessitated a fifth alarm to provide additional manpower for fire-fighting operations. A total of 20 engine companies and six ladder companies were utilized, not counting an engine company designated as the water supply company. The chief officers consisted of four battalion chiefs, one deputy, one assistant, a training officer, the Fire Marshal, the Chief of the Department, and the Fire Commissioner.

The fire, which was declared under control at 3:15 pm, resulted in heavy damage to the building and to the library collection. The roof had collapsed into the main reading room, where the contents were totally destroyed. The book stacks on either side were partially protected by the plaster and lath ceiling that fell on them. Books on the ends of the stacks toward the center of the main reading room received fire damage; books in the middle and the extreme outside ends of the stacks, smoke damage and water damage from hose streams. Some stacks, including the separate Lucas Hirst Collection, were protected with salvage covers placed by University personnel before the roof collapse.

[2] Temperature 91°F, relative humidity 41 per cent, wind southwest at 12 mph.

The library office area west of the masonry wall received smoke and water damage. Although the rolling steel fire doors failed to operate, Fire Department hose streams protected most of this area. The solid-core wood door between the cataloging department and the reading room was closed during the fire. This area received smoke and water damage, but the shelf list (the working card catalog used by the staff) was protected. Fire spread under the roof caused fire damage to second-floor offices above this area. Third-floor offices on the east side were gutted by fire.

Second-floor offices on the east side received varying degrees of fire damage. The offices in the northeast corner were more heavily damaged, with total burnout of the office adjacent to the reading room. The exposed lath and plaster wood-stud walls of this office were consumed as well as the ceiling above. The ceilings of three other offices were penetrated by fire and their contents received some fire damage.

The fire did not penetrate the basement area. Books stored in the basement stacks received water damage. Run-off water from hose streams accumulated to a depth of three feet. Books on the lower three shelves were wetted and the books above were dampened.

Salvage of the Library Collection. The salvage activities by University personnel and fire fighters during the fire were previously described. Around 4 pm that afternoon University personnel with Fire Department assistance were able to enter the building through the front entrance and salvage additional volumes. Approximately 20,000 volumes from the Lucas Hirst Collection were loaded on trucks and taken to another campus building. The card catalog used by the library staff, known as the shelf list, was later salvaged undamaged from the cataloging room.

The insurance carrier contacted two preservation experts from the Library of Congress to assist with salvage efforts at the library, and on Friday, July 28, these experts from the Preservation Office arrived at Temple University to oversee salvage operations. Another expert experienced in salvage techniques, from the American Philosophical Society Library in Philadelphia, also assisted. Up to this point the University had made no over-all salvage plans.

The first objective was to enter the basement stack area, assess the damage, and take steps to stabilize mold damage to the volumes. The warm summer weather coupled with the water situation in the basement meant that salvage work would have to be carried

Left: the scene at the rear of the building as the roof collapsed
PHILADELPHIA *Inquirer*

Top of page: exposure protection for Reber Hall

out promptly to control the damage to those books.[3] Volumes located in the basement were given first priority, as they were more difficult to replace, if not, as in the case of documents such as the Franklin Papers, irreplaceable.

The basement was opened on Friday at approximately 5 pm. Personnel were organized to provide emergency lighting and fans were also set up for ventilation. The labor situation in providing these emergency services was complicated by an on-campus labor dispute. Picket lines established before the fire by a local of an operating engineers' union hampered salvage efforts. By 2 pm Saturday a crew of volunteers had been organized to remove books from the basement. During the weekend 30,000 volumes were removed and placed in cold storage to retard mold growth. All the volumes were individually wrapped in freezer paper and packaged in boxes to be loaded on four 40-foot freezer vans for transportation to a cold storage plant. At the plant the books were frozen to a temperature of 20°F below zero. The remaining volumes in the basement were given eight sprayings of thymol to retard mold growth.

The volumes removed were mostly from the three bottom shelves, which had received the worst damage. The moldy, wet, and distorted books included many irreplaceable documents. The freezing operation had to be done quickly, to avoid further damage; thus the books were not coded or cataloged before packaging. At some later date the library staff will be able to remove the books from cold storage and decide which volumes are to be restored. This will depend on several factors, including salvage costs versus replacement costs, if it is possible to replace the volumes. Salvage of a legal collection is complicated, as many volumes are parts of sets. To have a volume missing from a set of books could reduce the value of the set.

The library collection also included a quantity of microfilm, all of which was saved by being placed in water and sent to the Eastman Kodak Company. The Kodak Company was able to reclaim all the microfilm for the University.

On Thursday of the following week the University decided to replace most of the volumes that had been housed on the first floor. Most of these volumes made up sets of reference books frequently used by students. A temporary facility with the replacement volumes — to be replaced by the new law library, scheduled for construction sometime early in 1973 — was needed for

the start of the Fall semester in September. The immediate replacement cost of approximately 50,000 volumes was estimated at $1 million.

That same day a decision was also made to salvage the remaining contents of the basement, about 120,000 volumes. Another crew of volunteers was organized for this task. The necessary resources were collected — including materials-handling equipment and boxes — and the crew was extensively briefed on procedures. The books were carefully indexed and boxed and loaded on trucks from conveyer belts. With this procedure approximately 5,000 books per hour were loaded. As mold growth had been stabilized with thymol, these books were transported to a controlled atmosphere storage facility.

Discussion. Philadelphia Fire Department officials conducted an investigation to determine the cause and origin of the fire. Investigators pinpointed the area of origin above the ceiling of a second-floor office at the northeast corner of the building. The probable cause was indicated to be of an electrical nature; but the electrical inspector's final report was not available at the time of this writing.

An investigation conducted by the insurance carrier centered around an abandoned stairway, also located in the northeast corner of the building, that ran between the street floor and the second floor. That investigation was continuing and further information on it was not available at the time of this writing.

Regardless of the exact cause, the fire apparently burned for some time within concealed spaces in the rear vacant office area. As was mentioned above, a contractor outside the building saw smoke coming from the roof. When the first Fire Department unit arrived, shortly after 1:54 pm, the cockloft below the roof in the northeast corner was heavily involved and the fire was progressing into the large-area concealed space below

[3] In general, leaving water-soaked materials in humid conditions at a temperature above 70°F for more than 48 hours can be disastrous, because of the growth of mold. See Peter Waters, *Emergency Procedures for Salvaging Flood or Water Damaged Library Materials* (available from the Preservation Office, Library of Congress, Washington, D.C.).

the main reading room. The Fire Department could not reach that area in time to prevent further spread – this in spite of aggressive initial interior attack and early roof-venting by two ladder companies with power tools. Fire-fighting operations were hampered by weather conditions and lack of full access for apparatus to the rear of the building. When fire fighters were overcome by heat exhaustion, a fifth alarm was needed to supply additional manpower. The water supply, from high-pressure and low-pressure mains, was plentiful.

The initial delayed discovery was significant to fire spread beyond the area of origin. By the time the alarm was telephoned to the Fire Department the magnitude of the fire was probably beyond interior attack capabilities. Investigators checked the fire extinguishers that security officers had attempted to use. These water-type extinguishers were found to be operative; apparently the security officers were not familiar with operating them. The nature of the fire spread, within wall and ceiling spaces, makes it doubtful that operation of the extinguishers would have been significant to the outcome of the fire.

The fire door failures in the front of the building were previously noted. The contents west of the masonry separation wall were protected from fire damage by salvage activities during the fire and by application of hose streams in that area. The one-hour-rated fire partitions above the suspended ceiling that had been required by building officials proved ineffective in preventing fire spread.

The sequence of events of this fire, with ultimate exposure to the library collection, is not unlike that of other fires in library facilities of ordinary construction where automatic extinguishing systems were not provided.[4] This fire could have been controlled by an automatic sprinkler system so arranged as to transmit an alarm to a central station or to the Fire Department. Sprinkler flow would certainly have caused less water damage than Fire Department hose streams.

[4] *Libraries,* FR60-1 (Boston: NFPA, 1961), 16 pp. Single-copy price, 50 cents.

The ultimate loss of contents will not be known until book restoration activities have been completed. Certainly the loss will exceed the $1 million replacement cost for references to establish a temporary library facility.

The successful salvage efforts will assist in reducing the contents loss. Library of Congress experts noted that the use of thymol in the basement was very effective in retarding the growth of mold. The thymol spraying also bought enough time to permit removal of the remaining basement collection. The freezing of 30,000 volumes from the basement will allow selective restoration of those materials at some future date.

It was noted that books that were tightly packed into shelves received less mold damage to individual leaves. By contrast, many of the books on the lower basement shelves swelled from water absorption and became distorted. Some of them fell from the shelves and exposed the interior of other books to mold spores.[5] Although books on the first floor were directly damaged by fire and hose stream discharge, there was less mold growth. The reasons for this were the ventilation conditions and the direct exposure to the atmosphere, which was less favorable to mold.

The University was very fortunate to salvage the working card catalog used by the staff, in that this record provided the key to salvage and restoration operations to rebuild the library collection.

This library loss indicates a need for preplanning for salvage and restoration of library collections. Prior knowledge of salvage techniques, a plan of action, and a listing of manpower and materials resources could have avoided the initial delays. Delays in salvage must be minimized to prevent mold damage to books and documents. △

[5] Library of Congress preservation experts indicate that mold growth developed within 36 hours in the basement collection. This rapid mold growth had a direct bearing on a decision to pump water from the basement the day after the fire. It might have been a better course of action to consult preservation experts prior to removing water from the basement. Alternative and immediate salvage measures could have prevented mold damage and avoided some restoration costs.

THE LAW LIBRARY FIRE: ANALYSIS

The fire at Klein Law Library revealed how rapidly fire can spread through a library building and showed that some of our conventional ideas about fire and fire protection may be in error. The sequence of events which follows is drawn directly from the preceding NFPA FIRE JOURNAL account of the fire:

Wednesday, July 25, 1972

1:44 PM	Smoke seen pouring out of the roof
1:45	Telephone alarm to the Philadelphia Fire Department
1:55	Six fire trucks on the scene; fire extinguishers fail to operate; building local alarm sounded to warn occupants of the fire; firemen attack fire at third floor level.
2:00	Heavy smoke; rolling steel fire doors in fire wall fail to operate.
2:05	Fire officer in charge sends second alarm, bringing four more engine companies and a ladder company; firemen on roof ventilate fire, cutting holes in walls, roof and ceiling, and a roaring fire is found between ceiling and roof.
2:30	Emergency fire brigade of eleven staff workers, thirty library patrons and some volunteers pass books and card files out of burning building; third alarm sounded, bringing another four engine companies and two more ladder companies; fire officer in command fears collapse of roof and orders everyone out.
2:35	Ceiling and roof collapse.
2:37	The fourth alarm is sent, bringing four more engine companies, one more ladder company; ladder pipes, elevated platform apparatus and hand lines combine to pour 11,000 gallons of water per minute into the burning building.
3:00	Water stands three feet deep in the basement stacks and rare book room; main reading room books badly damaged, those toward the center burned when the roof fell in, others soaked with water from hose streams; some shelves and the entire Lucas Hirst collection of 21,000 books in a separate first floor room are preserved by salvage covers placed over them early in the fire.
3:15	The fire is declared under control.

Analysis:

The Klein Law Library (building) was destroyed in spite of a number of factors which to the layman might have strongly suggested that it was a superior fire risk:

a. The library was situated in a very large city;

b. It was protected by an excellent fire department with almost unlimited resources in manpower and modern equipment;

c. The water supply was abundant and access for fire apparatus good;

d. Many people were in the building and on the adjacent streets; there was a maximum opportunity for someone to have had early knowledge of the fire and to turn in the alarm.

The occurrence of the Philadelphia disaster in spite of these favorable points should make the director of a library skeptical of giving credit to such things in making decisions about fire protection.

Elwood Willey of the National Fire Protection Association, who prepared the report of the Philadelphia fire for the NFPA FIRE JOURNAL commented:

> "The fire could have been controlled by an automatic sprinkler system so arranged as to transmit an alarm to a central station or to the fire department. Sprinkler flow would certainly have caused less damage than fire department hose streams."

A Single Sprinkler Head

This is an extremely conservative understatement. On the strength of the recorded sprinkler performance data for many thousands of fires, let it be said that the fire described would probably have been put out by action of a single sprinkler head[1]; further, the movement of water in the system would alert building engineers and the fire department at once, and the water would be shut off once the fire was under control. With advanced systems now available, it would probably have put out the fire and shut itself off before the engineer or the firemen got to the scene. Consider 20 or 30 gallons of water and negligible fire damage for the sprinkler-protected building as against the destruction of a building by fire and by the water damage from the necessary torrents of water thrown from hoses to put down the fire, 11,000 gallons per minute at one point.

This point needs to be made emphatically when fire protection for the library is being discussed, particularly an older, open plan library. The conclusions to be drawn from the Philadelphia fire experience strongly support automatic systems that will not only detect but extinguish fire. If incendiary fire strikes, especially, extinguishment must commence at once. Any delay, even for the call to the fire department and for their response, permits a fire to spread rapidly and means that the firemen may have to bring to bear large volumes of water to put it down.

The threat of damage to books from water has led to reluctance of architects and library directors to install automatic sprinkler systems. Proponents of sprinklers point to (a) the good experience of sprinklers in putting out library fires,[2] (b) the fact that water damage disasters in library fires have occurred only in libraries NOT PROTECTED WITH SPRINKLERS, (c) the fact that sprinklers now are available that stop putting out water as soon as they have put out the fire.

NFPA data[3] for the 40 years from 1925 to 1964 on automatic sprinkler operation shows that in 75,000 fires in all kinds of buildings protected by sprinkler systems 38 percent were controlled by a single sprinkler head opening over the fire and 71 percent by four heads at most.

The more recent experience[4] recorded by the National Automatic Sprinkler and Fire Control Association, Inc., indicated that 70 percent of all fires in sprinklered buildings are put down by action of a single sprinkler head, 13 percent by two heads and 10 percent by three heads. The risk of water damage incidental to a fire would seem to be under far better control with a complete sprinkler system than with even the most competent fire department, arriving after the fire is well underway.

REFERENCES

1. "History, Value and Performance Records of Sprinklers," Ch. 16, National Fire Protection Association Fire Protection Handbook, 13th Edition, 1969:16–11.
2. "Recommended Practices for the Protection of Library Collections," National Fire Protection Association, National Fire Codes, vol. 15, NFPA 910–1975:5,6.
3. "History, Value and Performance Record of Sprinklers," National Fire Protection Association Fire Protection Handbook, 13th Edition, 1969:16–11.
4. "Dry Pipe Sprinkler System," Bulletin 1.20, Automatic Sprinkler Corporation of America, Cleveland, Ohio, 1968.

Main Library, University of California at Berkeley

Alternatives for Protecting the Library Fire Risk

"The conflicting advice given to the architectural team, however, was a realistic reflection of the debate then taking place among archivists and librarians about the benefits and dangers of sprinklers. Apparently the case against sprinkler systems was presented more forcefully to the architects. When the planning of the actual structure was completed, sprinkler systems were missing from the design. The basis for the disaster twenty years later was beginning."

Walter W. Stender and Evans Walker
"The National Personnel Records Center Fire: A Study in Disaster," 1974.

Harry Elkins Widener Memorial Library, central unit of Harvard University Library

ALTERNATIVES FOR PROTECTING THE LIBRARY FIRE RISK

To weigh the alternatives in fire protection for a library and come to an informed decision the librarian or risk manager needs to have the advice of a fire protection professional. It is possible that the library has this kind of expertise within the organization, as is true of many large universities; however, insurance organizations concerned with a large fire risk usually want their own technical services specialists to survey it also.

The solution for the problem of the fire risk of an important library has a number of alternatives. Assuming that the library is lightly or poorly protected against fire, the decision maker may consider the following courses of action:

1. Accept the risk without change or improvement, based on a belief in the relatively low frequency of fires in libraries. He assumes that replacement of all essential collections would not be impossible.
2. Move the library to another building.
3. Build a new library building.
4. Improve the building to a superior kind of fire resistance by closing vertical openings (like open stairways), reducing too-large areas with fire walls and rated fire doors, and in other ways.
5. Install a modern automatic fire extinguishing system, properly engineered.
6. Divide the risk into fire divisions of manageable area (as in No. 4 above), and place choice collections in rare book areas under superior protection, such as a Halon 1301 total flooding system.
7. Protect the library with an approved early-warning fire detection system using products-of-combustion sensors and reporting an alarm to a central station and fire department.
8. Combine two or more of items (1) to (7).
9. Provide some lesser system of protection.

A One Percent Investment

The decision will likely be made on the basis of the cost of any protection program weighed against (a) importance of the library and the continuity of its services, (b) availability of funds, and (c) the informed judgment of the decision maker on the degree of need for improved protection. Generally speaking, the investment in a superior fire protection system for a large library structure should cost less than one percent of the total values at risk in structure and contents. Several fire systems may qualify for an insurance premium credit[1] sufficient to amortize a large part of the installation cost over a period of years. Leasing a system is an alternative to outright ownership.

Here are excerpts from an actual fire engineering report, recommendations for strengthening the fire safety of a university library which had some good construction features and some defects typical of older libraries:

1. Due to heavy fire loading, extra-high values, the presence of unprotected steel floor supports, and generally poor life safety aspects, an approved automatic sprinkler system should be installed throughout the library's stack levels. Both waterflow and control valve (electrical) supervision should be provided on the sprinkler system, tied into the campus police department and the city fire department.
2. The two open stairways in the stack levels should be totally enclosed to provide a two-hour fire protective enclosure. UL* listed 1½ hour Class "B" doors should be provided at the stair well openings.
3. The unprotected horizontal openings communicating from the library stack levels to the building floor areas should be provided with UL listed Class "A" automatic-closing fire doors. These doors may be maintained in the open position by utilizing ionization-type smoke detectors installed at ceiling level on both sides of the opening, interlocked with electromagnetic hold-open devices. Approved door closing and latching hardware should be provided.
4. UL listed manual fire alarm pull boxes should be provided and centrally located on all stack level floors. All boxes should transmit an alarm to the campus police department and the city fire department.
5. Lighted exit signs should be strategically positioned throughout all stack level floors showing direction to exits, including lighted signs over the exit proper. Illumination for the exit signs should be from a reliable source. In addition, emergency lighting should be arranged to maintain automatically a reasonable level of light at each sign in the event of failure of the normal source of power.
6. An approved Halon 1301 fire extinguishing system should be installed in the rare book collections room and vault on the 6th floor. This system can be monitored by the existing heat detectors.

What About Water Damage?

Will automatic sprinklers leak or discharge water when there is no fire? This is extremely unlikely, according to Bruce Harvey.[2]

> ". . . It is true that 'whether sprinkler systems may cause more harm than good' is a moot point with librarians. But this is due more to professional misconceptions than to any limitation of fire protection technology . . ."
>
> It should be self-evident, though the beliefs of librarians seem to indicate otherwise, that fire will always cause more damage to books than water. Combustion is a rapid, exothermic and irreversible reaction. Water dam-

* Underwriters Laboratories lists products which have been submitted for test and have passed required reliability testing.

age is slower, measured in hours or days, and gives time for the application of modern salvage techniques . . .

The possibility of sprinkler leakage is also cited as a drawback. Sprinklers rarely leak, the failure rate being approximately one in one million."

In the ALA project poll,[3] 153 libraries reported a total of 257 incidents of water damage; not one of these mentioned the action of automatic sprinklers, with or without a fire, as a source of water damage in the library. The eight categories cited were "broken water and steam pipes, water entering the building during construction, condensate, seepage, storm-driven rains, leaking roofs, faulty drains and sewers, and floods, i.e., rivers overflowing their banks."

Why do people sometimes think of automatic sprinklers in a fire as a source of water damage? The basis of this notion is probably their efficiency. When one or two sprinkler heads knock out a fire and then continue running until shut off, newspapers sometimes report that the fire did no damage at all, but there was "water damage" of this or that extent. Modern waterflow sprinkler alarms transmit an alarm electrically to a central station, so there is no longer any reason for any unnecessary water after a fire is put out, since help is summoned the minute water moves in the pipes, and someone can attend to the shutoff valve. With a modern on-off sprinkler system, the flow of water is stopped automatically.

Modern Developments

Recent years have seen the increasing acceptance of hydraulically calculated sprinkler systems, in which pipe sizes and sprinkler orifice sizes are calculated to meet the actual needs of the areas being protected, rather than being all of uniform size. Water fog sprinkler heads also can be used, which deliver a finely divided water mist, not only reducing the amount of water used but bringing it to the fire in a form which is the most heat-absorbent possible and the most efficient at stopping fire. These are developments which should be considered in the design of library fire protection systems.[4]

One obstruction to providing automatic sprinkler protection, particularly in an existing building, is the fear people have that sprinklers will destroy the appearance of a room or a building which has become widely known for its elegant decor. This is a problem for which there is an ever-growing family of solutions. Sprinkler heads are now turned out in almost every conceivable metal finish and color, as well as in several hundred configurations to serve special functions, so that it is not usually difficult to make them quite inconspicuous. Some heads are particularly designed so as to be invisible from below. Another invisible system is one installed above a suspended ceiling, the panels of which fall away in a fire situation and expose the sprinklers to the fire. This is an arrangement in use in portions of libraries at Massachusetts Institute of Technology.

No complete or "approved"* automatic sprinkler installation is known to have caused unnecessary water damage to books in library fires. Neither has accidental water damage been reported in libraries from such systems. A partial system does not provide complete protection. To illustrate this point, consider the fire loss at Ottawa in the Library of Parliament August 4, 1952.[5] Above a 3-story high book room and beneath a copper-covered dome was a concealed space which was not protected by sprinklers, an incomplete system which would not receive approval today.

An electrical fire in the dome, which could have been put down with one or two sprinkler heads, had instead to be attacked by firemen scaling the dome to a window 150 feet up. The sprinkler system kept fire away from the books, but about seven percent of the one million books and pamphlets were soaked by hose streams and sprinkler water. Because of good salvage work, very few items had to be replaced or discarded. For want of a complete sprinkler installation there was a loss of over $150,000.

Fire the Leading Threat

The Burns survey[6] of 1973 indicated that fire is the leading threat to library collections, and found that 38 percent of large libraries responding already have installed sprinklers. The National Archives building in Washington is protected by a standard wet pipe system,[7] and the new James Madison Memorial building of the Library of Congress will incorporate automatic on-off sprinkler heads on a wet pipe system, as well as other automatic systems.**

Other libraries upgrading fire protection have chosen automatic detection early warning systems, either alone or supplementing an extinguishing system. With products-of-combustion (POC) detection, very early knowledge of a fire is possible, alerting anyone at the building by an audible alarm signal, and by direct wire the 24-hour watch service at a central station. This may be the fire department; if not, the central station sents the fire message immediately on to the fire department. The alarm sounded in the building simultaneously makes it possible to locate the fire and take action to control it.

Oscar M. Trelles of the University of Toledo library made a study of fire protection and insurance for libraries[8] and found some good things to say about Halon 1301, the extinguishant used in modern systems for the protection of computers, electronic equipment and other things of extra-high value where water extinguishment might create a salvage problem. He describes an installation in Bentley Historical Library at University of Michigan. A library of over 57,000 cubic feet is protected with a system that produces a 6% concentration of Halon 1301 in air in a matter of seconds, sufficient to knock out a fire but not toxic to people, at least for short periods of exposure. The J. R.

* A system has to be approved by an insurance rating organization to qualify for reduction of premiums.

** Systems for detection and extinguishment of fire are described in Section 8, *Automatic Fire Protection Systems.*

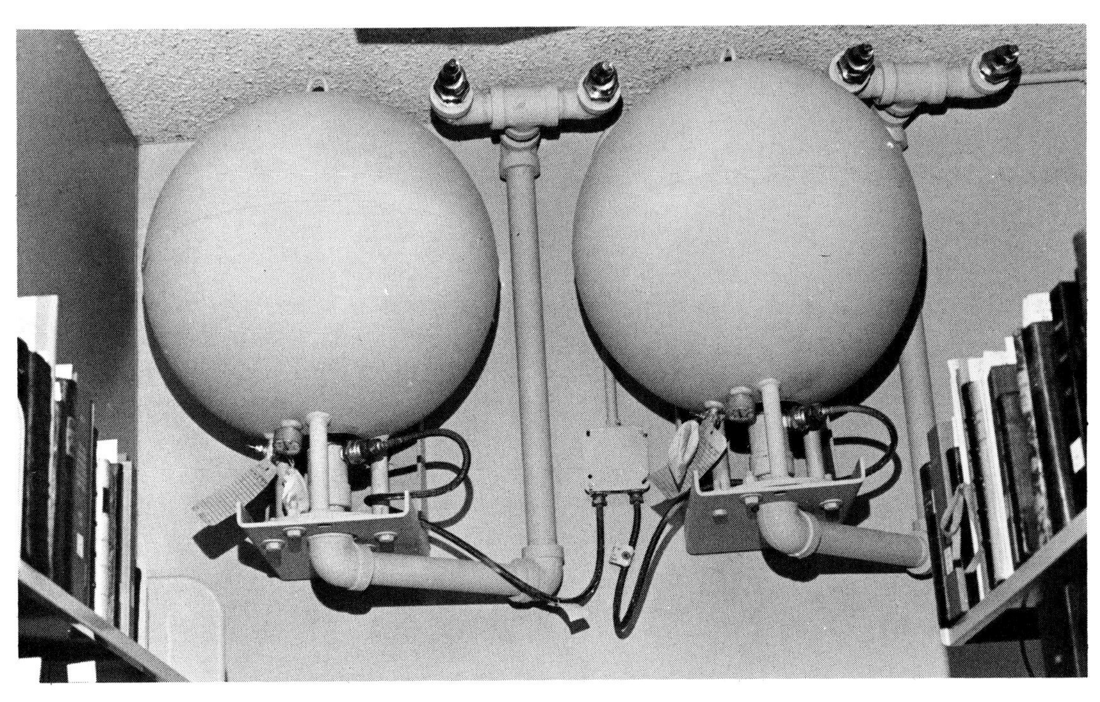

above:
A compact system; Halon 1301 protects a rare book area in the Shields Library at University of California, Davis.

left:

Kathleen Cifra, Asst. Department Head, Special Collections, demonstrates the manual trip lever for the Halon 1301 automatic system protecting rare books at Shields Library.

Blanchard Rare Books collection at the Shields Library, University of California at Davis is protected with a similar arrangement. (Other applications are mentioned in Section 8, Fire Protection Systems, along with new sprinklers using a minimum amount of water.)

In such a system the premature release of Halon 1301, which is somewhat expensive, is sometimes guarded against by cross-zoning detectors so as to require impulses from two systems before the agent is released. The delay is signaled by an audible alarm so that the staff can attack a nuisance fire (like one in a waste container) with an extinguisher and abort the automatic system manually, if need be. Another method, used at Winterthur museum in Delaware, provides a 5-minute delay for finding and suppressing the fire during working hours when the staff is present, following an alarm; this has worked well in an actual emergency.[9] With its special felicity for annihilating a fire without a trace, Halon 1301 is likely to be recognized by the library profession as the agent of choice for protecting any location where values are high and can be confined to a reasonably compact space.

REFERENCES

1. Fire Protection Handbook—NFPA—Revised 13th Ed. 1969:16–7.
2. "Sprinkler Systems and Books," a letter from Bruce K. Harvey, LIBRARY JOURNAL, vol. 99, no. 13, July 1974:1741.
3. "Protecting the Library and its Resources," American Library Assn., Chicago, 1963–4:18.
4. "Protecting the Library and its Resources," American Library Assn., Chicago, 1963–4:91–2.
5. NFPA 910, Recommended Practice for the Protection of Library Collections from Fire, 1970:5.
6. National Survey on Library Security, Burns Security Institute, Briarcliff Manor, N.Y., 1973:33.
7. Leo H. Swayne, "Fire Protection at the National Archives Bldg.," FIRE JOURNAL, vol. 69, no. 1, Jan. 1975:65–7.
8. Oscar M. Trelles, "Protection of Libraries," LAW LIBRARY JOURNAL, vol. 66, no. 3, August 1973:248.
9. Charles Ford, "Winterthur Revisited," FIRE JOURNAL, vol. 69, no. 1, Jan. 1975: 81–2.

Disaster Preparedness and Fire Prevention

"There is nothing the intellect resists more than a new idea. Step by step the logic leads us to the only practical solution; prepare for the fire before it happens."

Francis L. Brannigan
"Building Construction for the Fire Service," 1971.

Duplication of records and card files is important in disaster preparedness

DISASTER PREPAREDNESS AND FIRE PREVENTION

The occurrence of a fire or a major water casualty in a library quickly brings out furious activity and heroic efforts by volunteer labor to preserve materials and restore library services. There is a good deal of "hairtearing, false starts, confusion, communication problems, lack of information, general chaos . . ."[1] Some days later, with the restoration of the normal routine, there is an opportunity for establishing a formal program for preventing fires and being prepared for any emergency situation.

In Fremont, California, the Irvington branch library was destroyed in a fire commencing just before midnight, November 12, 1971. Frantic efforts were made to salvage wet collections, but there were numerous delays in getting authorization to act; it was impossible to get needed services and funds promptly. The damaged books finally were judged to be beyond recovery and were bulldozed into a sanitary landfill. Following this traumatic episode the Alameda County Library System[2] adopted a set of guidelines to avoid a recurrence, approximately as follows:

1. Copper wiring will be specified rather than any other type in county buildings.[3]
2. The maintenance, cleaning and adjustment of air conditioning and heating systems will be given a high priority.
3. Careful and frequent inspection will be made of electrical equipment and motors.
4. A monthly maintenance report will be required of branch libraries. All reported maintenance problems are to be given prompt attention—electrical deficiencies are to be repaired immediately.
5. Insurance plans are to be checked carefully and kept up to date to maintain adequate coverage.
6. Fire protection improvements will be made in the various branch libraries; automatic sprinklers and automatic products-of-combustion detection systems will be planned and budgeted.
7. Fire departments will be closely involved in library disaster planning and invited to make company inspections in the libraries. They will hold keys to branch libraries and lists of telephone numbers to alert library staff people to any emergency.
8. A policy of urgency is adopted for any fire emergency, calling for immediate action to salvage collections and remove them from the fire scene; there must be no delay, even for investigators and adjustors, which would delay salvage operations.

The new policy was put to the test when an arsonist set fire to the loan desk area of the Centerville Branch Library June 1, 1974 during the night. Water damage was held down as the fire department made an attack on the fire from a side entrance so that hose streams were not directed toward bookshelves. Consequently many books exposed to the fire were affected primarily by soot and smoke, rather than damaged by water. The emergency plan brought ten library people promptly to the scene ready to work. Books were wiped off, packed into boxes and removed from the scene. In 36 hours the salvage operation was complete and 70 percent of the collections had been removed. Fire loss to the library was estimated at $20,000. A fire sale was held and about 8,000 slightly damaged books were sold.

Water Emergency Guidelines

There are various casualties other than fire which subject books and libraries to water damage, such as flooding, broken pipes or other mechanical failure, and extraordinary storms. Water casualties were under discussion in the special disaster session of the California Library Association at San Francisco in December, 1973.

Mills College had suffered water damage incidental to failure of a slip joint washer in a hot water line to the book binding studio, with the result that water leaked on periodicals, including some fine publications with color plates. As an afterthought following this casualty, several points were mentioned by Diana Thomas, librarian, as essential in library disaster planning for the water emergency:

1. Electric pumps may be needed; it is important to find out where they can be borrowed promptly.
2. The electrocution hazard of control panels (and any electrical apparatus) in wet areas must be kept in mind.
3. The possibility of water damage should be considered in planning the layout and mechanical equipment of the library.
4. Know in advance where freezers can be obtained on short notice for emergency storage of wet books. (Mills College found a benefactor in their hour of need when an ice cream plant provided refrigerated trucks for a temporary cold storage for wet library materials.)

Disaster planning at Cornell University Libraries[4] following their flood in 1972 developed these memoranda:

1. Designate (in advance) as salvage officers any staff members with experience in recovery of wet materials.
2. Have a written plan.
3. Locate sources of supply for portable dehumidifiers.
4. Have a list of workers or a source for recruiting volunteers promptly.
5. Know toxic effects of any salvage materials and control measures needed.
6. Plan how and where to handle a possible emergency of wet books.
7. Educate staff in details of the plan.

8. Set up a Committee on Safety and Emergencies, to have such responsibilities as the following:
   —Make plans and preparations for all possible disasters;
   —Arrange for fire training and first aid training;
   —Coordinate efforts with other university departments;
   —Appoint an emergency coordinator for each library;
   —Hold "postmortems" of actual emergencies;
   —Prepare an emergency evacuation plan;
   —Inspect emergency equipment; "ferret and report" hazards to safety, using a detailed checklist.
   —Prepare a manual for emergency situations and recovery procedures.

Fire Prevention

A *fire prevention* program for the library *as a substitute for structural soundness and automatic fire protection systems* falls short of being adequate protection. However diligent the housekeeping, the smoking discipline, the fire emergency drills, there are situations which will periodically produce destructive fires. Prof. Francis L. Brannigan, the Maryland fire science educator, used to tell his students *"Fire prevention is for kids."* In other words, "When high values are at stake, never rely on fire prevention routines alone; you have too much to lose!"

Nevertheless, fire prevention and loss control are worthwhile conservation programs that should be applied in every library, large and small. Here are some principles that should be observed:

1. Library staff should become thoroughly familiar with
   —The function of any automatic fire systems;
   —The fire extinguishers—how to use them;
   —The annunciator panel, if any, and how to read the location of a fire from it;
   —The importance of standard fire prevention measures, good housekeeping, smoking discipline, etc.
2. Location of sprinkler valves must be known to permit closing valves when a fire has been put out.
   —Sprinkler plugs can be obtained and building employees trained in placing them into a discharging sprinkler as an alternative to closing down a whole branch of the system after a fire is out (applies to standard wet pipe systems).
3. Essential files, including the library's shelf list, should be duplicated and one set maintained in a separate building.
4. Contractor's operations should be very carefully watched; contractors' employees have been responsible for many bad fires.*

* Fires of this origin described in Section 1 were the Birmingham, Wheaton College and University of Texas fires.

—Notify the fire department or fire marshal about contractors commencing jobs in the library.
—Demand a fire watcher for welding and cutting operations or other "hot work."
—Obtain a description of measures being taken by the contractor to avoid fires.

5. The fire department concerned should be invited to become fully familiar with the library and its fire systems.
—Fire department officers should serve on the disaster committee or at least attend early planning sessions.
—Inspections should be made by the fire department in the library to familiarize them with the building.
—Periodic fire inspections should be requested of the fire department or fire prevention bureau.

Security and fire safety of the library generally, and particularly of rare books and special collections, should be under constant study.

REFERENCES

1. David Y. Sellers and Richard Strassberg, "Anatomy of a Library Emergency," LIBRARY JOURNAL, vol. 98, no. 17, October 1, 1973:2824–7.
2. From an interview with Caroline Long, Management Analyst, Alameda County Library System, Hayward, Calif., November 4, 1974.
3. "Correcting Aluminum Wiring Troubles . . .", FIRE JOURNAL, NFPA, vol. 68, no. 4, July 1974:37.

VII.

Salvage of Wet Books

"The first step is to establish the character and degree of damage. Once an accurate assessment of the damage has been made, firm priorities and plans for salvaging the damaged materials can be drawn up. These plans must include a determination of the special facilities and equipment required. Overcautious, unrealistic, or inadequate appraisals of damage can result in the loss of valuable materials. Speed is of the utmost importance, but careful planning is equally essential in the salvage effort.

Peter Waters
"Procedures for Salvage of Water-Damaged Library Materials," 1975.

Entrance Hall of University Research Library, U.C.L.A.

SALVAGE OF WET BOOKS

Everyone agrees that books should not be wet, and that the only possible excuse for throwing water on them is to put out a fire. When fire can be controlled by automatic sprinkler action or by immediate action of a manual fire extinguisher or a small hose, the fire damage is limited and a moderate amount of water controls it. When a fire is in progress for some time before it is discovered and attacked, and it has to be controlled by a fire department using hoses, the water damage may be very extensive. It is because of damage from fire hoses and floods that the science of salvage of books has been developed.

Fire destroyed a regional library at Gothab, Greenland, February 9, 1968. Books and other materials wet by hose streams quickly froze. They were shipped to Denmark frozen and placed in cold storage, from which books were gradually brought out a few at a time and restored by conventional air drying.

Other items awaiting salvage were more fragile, such as handwritten letters, manuscripts and maps of the missionary Samuel Kleinschmidt, over 100 years old. These were sure to be damaged if air dried, since the ink would run. They were set aside for freeze-drying, and the Food Technology Laboratory of the Technical University of Denmark was requested to undertake the salvage.

After some experiments had been done to develop a technique, materials were placed in heated vacuum chambers for periods of one and one-half to two days to reduce the water content, and the vacuum gradually released. The process was entirely successful for all of the hand-written documents. "Each page separated easily and in no case did the ink run."[1] One small failure resulted when the vacuum chamber cycle was not quite long enough for a thick photo album in which the ice persisted; the melting caused the ink to run, showing that the freeze-drying process was indeed necessary.

A Book 100 Years Under Water

In another application of freeze-drying using the vacuum chamber, a book that had rested in deep water for nearly 100 years was restored. This was a copy of the *Merchant's Almanac* which had lain in the wreck of the steamer *Bertrand* since 1865 in the Missouri River. The book was placed in a commercial deep-freeze; from there it was taken to the Smithsonian Institution and put into a vacuum chamber at –20°C and the vacuum drawn down to 150 microns. By removing the book before it was completely dehydrated, it remained flexible rather than brittle. "The end result was a well-preserved book with pages that separated easily."[2]

Two massive salvage efforts which recently advanced the science of restoration of water-soaked books and papers were the result of disastrous fires in buildings not protected by sprinkler systems, into which fire companies were obliged to pump torrents of water. These fires destroyed the Charles Klein Law

Library[3] in Philadelphia, July 25, 1972, and a large part of the National Military Personnel Records Center at Overland, Missouri on July 12, 1973.[4]

Following the Klein Law Library fire in Philadelphia, the first process tried for salvage of thousands of wet books was freeze-drying. This was adequate for individual volumes, but did not work with large numbers of books—sixty or seventy at a time. However, it was determined that freezing is usually a good and necessary intermediate step between the wetting of books and their eventual recovery; freezing stabilizes the materials to prevent further deterioration. Wet books were dipped in a thymol solution to retard mold, wrapped in paper and then frozen at –20°C (–4°F).*

The books were then taken to General Electric Space Simulation facilities at King of Prussia, Pennsylvania, near Valley Forge, and placed in vacuum chambers. Air was exhausted from the chambers until an altitude of 100,000 feet was simulated. Shelves were then heated to slightly more than 100°F so as to boil off water into vapor without damage to the paper. A refrigerator coil then collected the moisture by condensing it directly from vapor.

Each cycle accommodated six to seven thousand books and a cycle lasted for one week. Cost of salvage was estimated at $1.80 to $1.90 per book at the Space Chambers. However transportation, cold storage and cleaning costs would have to be included in figuring the total cost of restoration.

George A. Reese, Jr., Insurance Manager for Temple University, recommends[6] keeping additional data and duplicating shelf lists so that total destruction of the principal shelf lists would not make it impossible to determine losses and make replacements. The duplicated data should, of course, be stored in another building under adequate protection. Another recommendation he makes is to reduce the cost of recovery of wet books by first screening out junk material and books too badly damaged to be salvaged.

He mentions such problems as:

(a) Recurrence of mold in some restored books;

(b) Need for rebinding of many salvaged books;

(c) Need for professional cleaning and better methods of cleaning. Many methods were used, including painstaking work with the plastic or putty variety typewriter typecleaner. There was no single "best" method.

Problems of the fire, water and salvage emergency were numerous for Prof. Erwin Surrency, librarian of the Klein Law Library, for whom the fire was a wholly traumatic episode. He lost such collections as the Benjamin Franklin imprints of the laws of Pennsylvania and a very large collection of the known justice of the peace manuals, some of them very rare, as well as rare books about famous trials in England and America.

His most chilling memory of the disaster is that persons in the burning building were in real danger of being trapped in the fire. His advice to fellow librarians:[7]

> "What have I learned from this fire? I guess I have learned quite a lot. First of all, I would say keep better records. Somehow our cataloguing practices were inadequate for an accurate record of what the library

* Current doctrine specifies –20° F [5]

possessed. Many of us do not keep records of serials, so we do not have an item count of precisely the materials we have.

Secondly, my advice is to squawk when the fire marshal calls your attention to the fact that the building is in direct violation of the fire code. Unfortunately, our building had violated the fire code in several respects. I would certainly urge all librarians to make sure their buildings conform to the fire code."

Salvage operations for wet materials following the Military Records Center fire were similar to the Klein Library experience. The principal difference was that much of the material was taken directly to vacuum chambers and dried without being first frozen. This is somewhat quicker and cheaper than freeze-drying. In freeze-drying water in the books is caused to sublime, or pass directly from the frozen state to vapor without going through the liquid state. Freezing not only stabilizes water-soluble inks and dyes, but prevents mold, reduces stains, eliminates smoke odor, and stabilizes damage for whatever period is needed to arrange for drying. Wet materials have been held in the frozen state without deterioration for as long as six years.[8]

Evans Walker[4] says it is not necessary to freeze-dry books if they can be taken directly to a vacuum chamber before mold sets in. Some of the Military Records salvage work was done with materials that had been wet for four months and were very moldy. Sterilizing the books to kill the mold would have been too expensive. The vacuum chamber process arrested the development of the mold, and the decision was made to provide air-conditioned spaces for storage of the restored materials rather than go into the cost of sterilizing to prevent recurrence of mold.

The McDonnell-Douglas Space Systems Laboratory is credited by Mr. Walker for special efforts in many ways to make the Missouri Records Center salvage successful. The first load of books was returned too dry, after which the drying cycle was cut from five days to four, and moisture was added at the end of the final cycle to restore the proper level of humidity. Books too dry become brittle.

There were four chambers used, three at McDonnell-Douglas of 2,000 and 1,000 and 1,500 cubic foot capacity, the latter a cylindrical chamber in which books were suspended in nets. The fourth facility, a very large vacuum chamber, was then discovered at a Sandusky, Ohio NASA facility. Many of the remaining loads were taken there directly by truck.

Since that time McDonnell-Douglas has been able to achieve an excellent restoration of architectural drawings soaked in a flooded Chicago basement. Some of these were unique and highly prized items, the original work of the master architect, Louis Sullivan.

Emergency planning for any important library should include not only the possibility of fire, but the restoration of wet books and papers after a fire, or after any other casualty in which books might be thoroughly soaked.

Water casualties from other sources than firefighting have been fairly common in recent years. Floods destroyed or caused severe water damage to many

New York and Pennsylvania libraries in 1972;[9] there were also numerous minor incidents. A ball float malfunctioned in the attic above the Map and Geography library at University of Illinois in Champaign,[10] wetting and spattering many books, all but two of which were restored. In a similar incident, pressure tank equipment designed to *prevent* water damage in the Library Company of Philadelphia[11] building discharged from two to five inches of water on five stack levels, damaging elevator machinery, catalog cards and periodicals, but no books.

Consultant on both the Klein Law Library and Military Records Center salvage operations was Peter Waters, Restoration Officer of the Library of Congress. He had previously served as conservator in the British Museum, and had an important part in salvage efforts following the floods in Florence in 1966. His treatise, called "Procedures for the Salvage of Water-Damaged Library Materials," was revised by the author and republished by the Library of Congress early in 1975. This is the definitive guide for any library in the water damage emergency.

Another publication on salvage operations is that of the Corning Museum of Glass, based on their disastrous flood of June 23, 1972. All flooded materials were promptly frozen to prevent deterioration. A program of research in conservation techniques was established and David J. Fischer, Ph.D., was commissioned to carry out the program and prepare research reports, which form a considerable part of their book, due for publication in 1976.

REFERENCES

1. James Flink and Henrik Hoyer, "Conservation of Water-Damaged Written Documents by Freeze-Drying," NATURE, vol. 234, December 17, 1971.
2. R. O. Hower, "Advances in Freeze-Dry Preservation of Biological Specimens," CURATOR, vol. XIII, no. 2, 1970:135–152.
3. A. Elwood Willey, "The Charles Klein Law Library Fire," NFPA FIRE JOURNAL, vol. 66, no. 6, Nov. 1972:16–22.
4. Evans Walker, "Military Personnel Records Center Fire: Part 2, Records Recovery (Salvage of Wet Papers)," NFPA FIRE JOURNAL, vol. 68, no. 4, July 1974:65.
5. Peter Waters, "Procedures for the Salvage of Water-Damaged Library Materials," Library of Congress, Washington, 1975:7.
6. Conversation with George A. Reese, Jr.
7. Erwin C. Surrency, Moderator, panel discussion "Guarding Against Disaster," LAW LIBRARY JOURNAL, vol. 66, no. 4, November 1973:420.
8. Peter Waters, "Procedures for the Salvage of Water-Damaged Library Materials," Library of Congress, Washington, 1975:6.
9. "A Visit to Flood Damaged Libraries," American Libraries, vol. 3, no. 8, Sept. 1972:845–6.
10. "Illinois Volunteers Save Soggy Books," LIBRARY JOURNAL, vol. 98, no. 18, Oct. 15, 1973:2955.
11. "Water Damage at the Library Company of Philadelphia," LIBRARY JOURNAL, vol. 94, no. 12, June 15, 1969:2404.

Automatic Fire Protection Systems

"The removal or lessening of hazards will not stop fires altogether. The question of fire detection, alarm and extinguishing must be considered."

Keyes D. Metcalf
"Planning Academic and Research Library Buildings," 1965.

Library of Congress: James Madison Memorial Building

The new James Madison Memorial Building of the Library of Congress has vast floor areas, nearly 8 acres on each of 3 levels below the main entrance level and over 4 acres on each of the upper floors. It is protected against fire with a modern complex of systems for detection, warning and suppression. It is the largest of several great new libraries with fire protection design developed through the systems approach.

AUTOMATIC FIRE PROTECTION SYSTEMS

The presentation of fire protection systems and devices in this section is intended to be readable for non-engineering professional people. It is not intended to recommend one device or system above another, except that CO_2 total flooding systems are specifically not recommended for libraries. Generally speaking, the *listing* by Underwriters' Laboratories or *approval* by Factory Mutual conveys evidence of reliability and durability according to testing procedures of those organizations. Almost all good fire protection equipment has this kind of approval, and codes usually demand "listed" or "approved" devices or systems when systems are required. Technical information is available from manufacturers and from such standard references as the following:

1. *Best's Safety Directory,* A. B. Best Co., Morristown, N.J. 07960.
2. *Fire Protection Equipment List* (annually), Underwriters' Laboratories, 207 E. Ohio St., Chicago, Ill. 60611.
3. *Fire Protection Handbook,* 14th Edition, 1976, National Fire Protection Association 470 Atlantic Avenue, Boston, Mass., 02210.
4. *Protecting the Library and Its Resources,* 1964, American Library Association, Chicago, Ill.
5. *Approval Guide* (annually,) *Equipment, Materials, Services for Conservation of Property;* Factory Mutual Engineering Corp., 1151 Boston-Providence Turnpike, Norwood, Mass. 02062.

When a superior level of fire protection of collections is needed in the library, automatic fire systems are clearly indicated. Systems are of two general types according to function, *automatic fire detection* and *automatic extinguishment.*

The summary below lists in brief form both types, commencing with basic early warning heat sensing systems.

(A) *Automatic Detection Systems:*

1. Automatic fire detection using fixed temperature sensors;
2. Automatic fire detection, using rate-of-rise sensors which report a sudden increase of heat;
3. Automatic smoke detection, reporting an alarm when visible smoke obstructs a light source in the sensor;
4. Automatic products of combustion (POC) detection, using sensors which recognize invisible products of combustion even before smoke appears, and sound an alarm.

(B) *Extinguishing Systems:*

1. Automatic sprinkler system: pipes with nozzles (or heads) having soft metal components which fuse (melt) at a pre-set temperature and spray water on a fire;

2. Dry pipe automatic sprinkler system; same as above, but with air, not water, under pressure in branch piping until water is demanded by fusing of a head;
3. Pre-action automatic sprinkler system; pipes are empty until water is introduced by action of rate-of-rise detector; heads then fuse individually as in the standard wet pipe system;
4. Pre-action *"Firecycle"* automatic sprinkler system; similar to #3, except that the valve controlling the branch line water supply closes automatically when the fire is out, stopping the flow of water, and reopens if necessary;
5. *Aquamatic* stop-and-go sprinkler heads; usually installed on a wet pipe system. Each head opens over a fire independently on action of a snap disc at the head, and closes when fire is out, as often as necessary;
6. Halon 1301 system; puts out fire quickly through action of a compressed vapor which is released through nozzles, responding to any standard means of detection;
7. High-expansion foam, containing little water, is used in airplane hangars and industrial buildings. It has a potential value for library protection, but has apparently not yet been so used;
8. CO_2 (carbon dioxide) used as a total flooding system presents a suffocation problem; it is not recommended.

For all the above systems it should be said that the action of the system should also send automatically an alarm to a central station, either a private alarm center, a police communications post or the fire department directly. A local alarm is usually wanted also; anyone near an incipient fire may be in no immediate danger, and may be able to take some intelligent action to put out the fire or confine it. However, a system that provides the local alarm only has limited value. Even a waterflow sprinkler gong alarm on the outside of a building will excite only a mild curiosity on the part of the average person passing by; there must be automatically an urgent message at once to a fire department or another with a direct responsibility for protection of the library, and with a faculty for being alert at all hours.

The pages which follow show drawings and descriptions by manufacturers of some of the most common fire protection devices and systems. These materials were selected for clarity and the absence of extravagant claims. The fact that something is shown here does not mean that it is recommended, or endorsed, or even used in library fire protection. Information and guidance on these matters can be obtained from fire protection professionals.

1. AUTOMATIC SPRINKLERS—WET PIPE SYSTEM

Action

1. Sprinkler heads are on pipes charged with water and designed to open over a fire at 165°, 212°, 286° or 360° degrees Fahrenheit.
2. Heads open individually.
3. When fire is out, water is shut off by closing a valve supplying a branch line.

Comments

1. Notable for economy and simplicity of installation, maintenance and repair.
2. Proved in daily use since 1874.
3. Acceptable hardware includes that listed by Underwriters' Laboratories or approved by Factory Mutual laboratories.
4. Most numerous of all systems.
5. Has a formidable reputation for preventing loss.

A general description of wet pipe sprinklers is that of one of several manufacturers, on the two pages following.

WET PIPE SPRINKLER SYSTEM

Instant Protection — Easy Maintenance

FOR ALMOST ALL TYPES OF BUILDINGS IN NON-FREEZING AREAS

The Wet Pipe Sprinkler System, a system that meets most fire protection requirements, is also one of the simplest and most widely used.

This rapid, reliable system is used for normal hazards, indoor or out, where temperatures do not fall below freezing.

Water is right at the sprinkler, ready to start fighting fire the instant a sprinkler opens. Temperature-rated fusible-link sprinklers prevent water discharge in areas not affected by the fire.

The Wet Pipe Sprinkler System extinguishes or controls fires with a minimum of sprinklers operating. The graph, at right, illustrates the number of sprinklers that were necessary to extinguish fires under every conceivable fire situation, covering a ten-year period. These are averages including all types of buildings.

REPORT OF NATIONAL AUTOMATIC SPRINKLER AND FIRE CONTROL ASSOCIATION, INC.

Typical "Automatic" Sprinkler Wet Pipe System applications:

BAKERIES · BREWERIES AND BOTTLING WORKS · CANNERIES · COTTON AND WOOLEN MILLS · DAIRIES · FEED MILLS · FOUNDRIES · GRAIN ELEVATORS · LAUNDRIES · MERCANTILES · METAL WORKING · PAPER MILLS · RESTAURANTS · STEEL MILLS · WAREHOUSES · APARTMENTS · CHURCHES · HOSPITALS · CARE HOMES · HOTELS · OFFICE BUILDINGS · PUBLIC BUILDINGS · SCHOOLS · THEATERS

"Automatic" Sprinkler CORPORATION OF AMERICA • CLEVELAND, OHIO 44141

OPERATION

The piping of a Wet Pipe System is filled with water, under supplied pressure, at all times. A hinged clapper in the Alarm Valve keeps the water in the piping and right behind every sprinkler, where it is ready to go into action.

When there's a fire, the fixed temperature sprinklers fuse and water instantly discharges onto the fire. The water flows into the sprinkler piping and through only those sprinklers that are fused by the fire. At the same time, a water motor fire alarm or electric fire alarm sounds. The "Automatic" Sprinkler Wet Pipe System is UL and FOC listed, FM approved.

To assure that the water supply is not accidentally shut off, an optional monitor switch may be installed on the water control valve and connected to the electric alarm to serve as a supervisory device. Any movement of the valve actuates a switch which sounds an alarm.

A nonflammable anti-freeze solution may be used in place of water in the sprinkler piping of a Wet Pipe System used to protect a small area. This will assure protection at below freezing temperatures.

The Wet Pipe Valve is rugged cast iron—flanged and designed for maximum working pressure of 175 PSI. 4″, 6″ and 8″ sizes.

In areas where the temperature is likely to fall below 32 degrees, an "Automatic" Dry Pipe System may be used, indoors or out. For a complete explanation of the system, see Bulletin 1.20.

And, for detailed information on "Automatic" Sprinklers—temperature ratings, spray patterns and finishes—and, accessories, refer to specific bulletins.

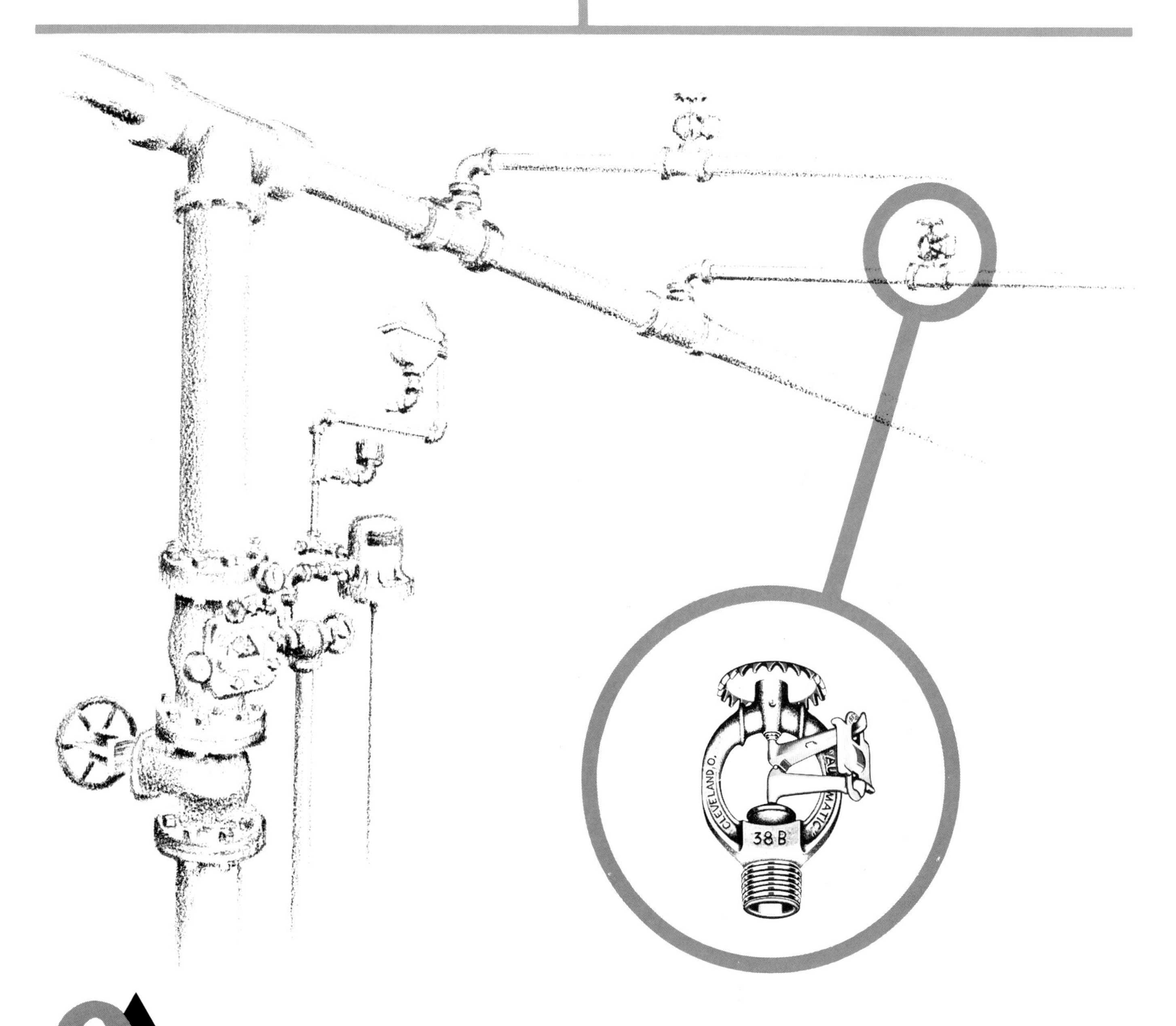

"Automatic" Sprinkler CORPORATION OF AMERICA • CLEVELAND, OHIO 44141

2. AUTOMATIC SPRINKLERS—DRY PIPE SYSTEM

Action

1. Sprinkler heads are on pipes charged with air and designed to open over a fire at 165°, 212°, 286° or 360° degrees Fahrenheit.
2. Heads open individually.
3. Air is vented from the open head or heads, and water then advances through the pipe and is discharged at the same openings.
4. When the fire is out, water is shut off at a valve supplying the branch line.

Comments

1. Used primarily where sprinklers are needed in unheated areas in cold climate.
2. Action is somewhat slower than wet pipe system and more heads may be expected to open over a fire.*

* Fire Protection Handbook, NFPA, 13th Ed. 1969:16–10

A general description of dry pipe sprinklers is that of one of several manufacturers of dry pipe hardware, on the two pages following.

DRY PIPE SPRINKLER SYSTEM

Dependable Protection

FOR ALMOST ALL TYPES OF BUILDINGS, IN ANY TEMPERATURE

The Dry Pipe Sprinkler System is one of the most effective of the basic sprinkler systems. It may be used in almost any area, at any temperature—above or below freezing.

Air under pressure, not water, is in the sprinkler piping. Air is maintained in the system until heat from a fire fuses one or more of the temperature-rated, fusible-link sprinklers. Water then enters the entire piping system to be discharged only through the spinklers that have been fused by fire.

The Dry Pipe Sprinkler System extinguishes or controls fires with a minimum number of sprinklers operating.

REPORT OF NATIONAL AUTOMATIC SPRINKLER AND FIRE CONTROL ASSOCIATION, INC.

Typical "Automatic" Sprinkler Dry Pipe System applications:
UNHEATED BUILDINGS AND STRUCTURES OF ALMOST ANY TYPE · COLD STORAGE WAREHOUSES · LOADING DOCKS · PARKING RAMPS · OUTDOOR PIERS · OUTDOOR PLATFORMS

OPERATION

The piping of a Dry Pipe System is filled with air that is kept under pressure and retained in the sprinkler piping by the fusible-link sprinklers. The air pressure pushes against a hinged clapper in the Dry Pipe Valve and seals off the water supply in the main riser. The valve is housed in a heated area so it can't freeze up.

When there's a fire, the fixed temperature sprinklers fuse and the air pressure decreases. The water supply then pushes open the clapper, flows into the sprinkler piping and through only those sprinklers that are fused by the fire. At the same time, a water motor fire alarm gong or electric fire alarm sounds. The "Automatic" Sprinkler Dry Pipe System is UL and FOC listed, FM approved.

To assure that the water supply is not accidentally shut off, an optional monitor switch may be installed on the water control valve and connected to the electric alarm to serve as a supervisory device. Any movement of the valve actuates a switch which sounds an alarm.

The Dry Pipe Valve is rugged cast iron—flanged and designed for maximum working pressure of 175 PSI. 6" size.

In areas where the temperature will **not** fall below freezing, an "Automatic" Wet Pipe System may be used. For a complete explanation of the system, see Bulletin 1.10.

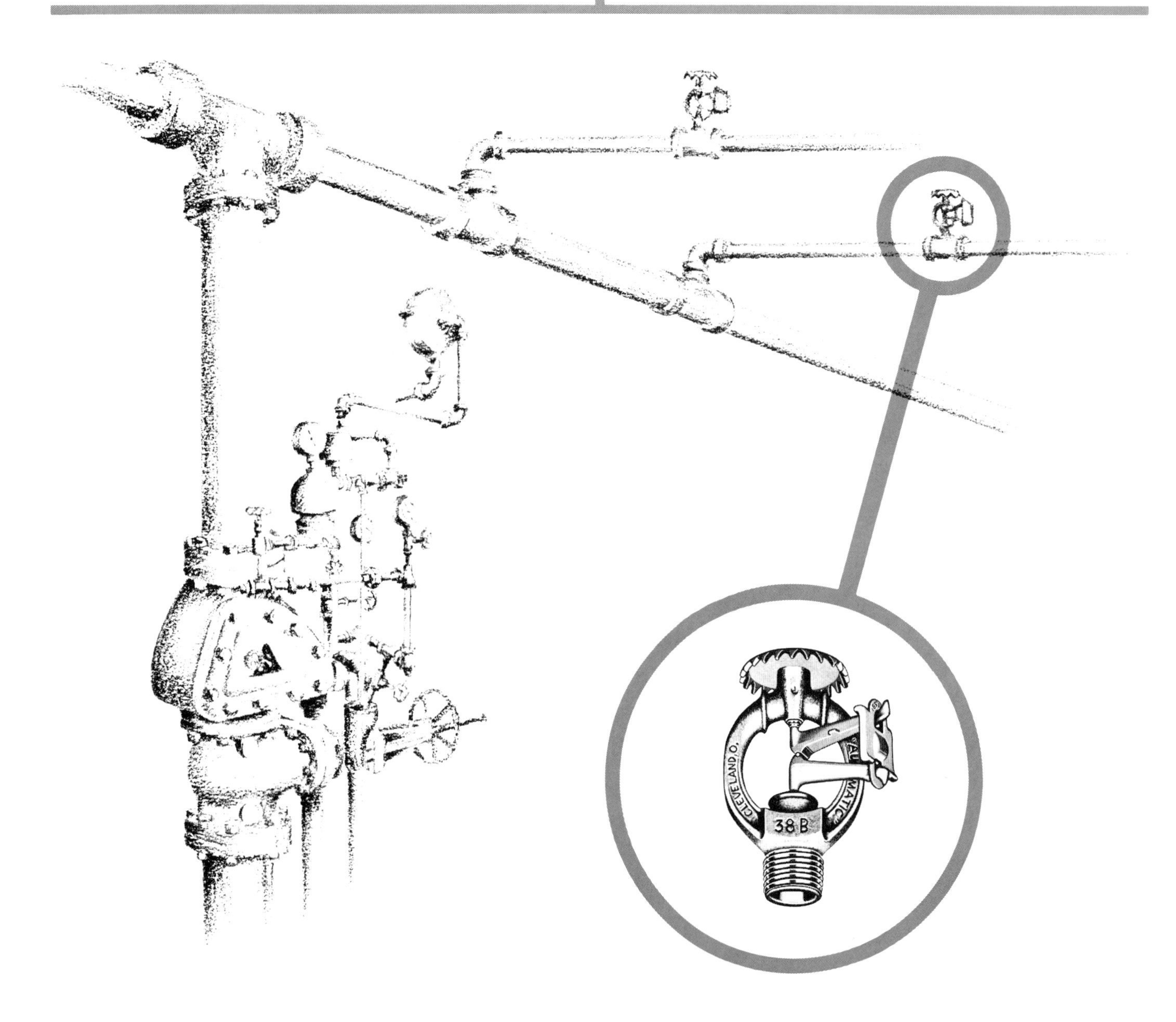

3. AUTOMATIC SPRINKLERS—PRE-ACTION SYSTEMS

Action

1. Empty pipe branch lines become charged with water when an automatic detection system senses abnormal heat rise.
2. Sprinkler heads are on pipes charged with water and designed to open over a fire (usually) at 165°, 212°, 286°, or 360° degrees Fahrenheit.
3. Heads open individually.
4. When fire is out, water is shut off by closing a valve supplying a branch line.

Comments

1. Reduces further the remote possibility of accidental and unwanted discharge of water through damage to sprinkler pipes or heads.

A typical pre-action sprinkler system is discussed in the bulletin on the three pages following.

SUPROTEX PRE-ACTION SPRINKLER SYSTEM

Rate-of-Rise Activated—Fusible-Link Sprinklers

FAST, AUTOMATIC ACTION IN ALARM, CONTROL AND EXTINGUISHMENT

The SUPROTEX Pre-Action Sprinkler System is specially designed to provide superior fire protection without unnecessary water discharge in areas of high-value content. It is installed in heated or unheated structures with light, ordinary or extra-hazard ratings —where early alarm is needed for evacuation of personnel and notification of fire-fighting agencies.

The SUPROTEX Pre-Action Sprinkler System combines fast, dependable Rate-of-Temperature-Rise detection and fusible-link sprinklers.

For a complete explanation of Rate-of-Rise detection, see Catalog Sheet 2.00. The SUPROTEX-DELUGE Sprinkler System, with open-head sprinklers that flood an entire area simultaneously, is described in catalog Sheet 2.10.

ADVANTAGES

SUPROTEX Pre-Action Sprinkler Systems provide many flexible advantages, including:

1. Rapid detection of fire.
2. Fire alarm sounds **before** water discharge.
3. Allows time to extinguish small fires manually before sprinklers go into action.
4. Can be used in both heated and unheated structures.
5. Eliminates accidental water discharge if sprinklers or piping are damaged.
6. Maintains constant protection, even while sprinklers or piping are being serviced.
7. Sounds fire alarm and stands ready to extinguish fire if detection system is damaged.
8. Sounds trouble alarm if sprinklers or piping are damaged.
9. Sounds trouble alarm if main control valve is not fully open.

BASIC SUPROTEX PRE-ACTION SPRINKLER SYSTEM

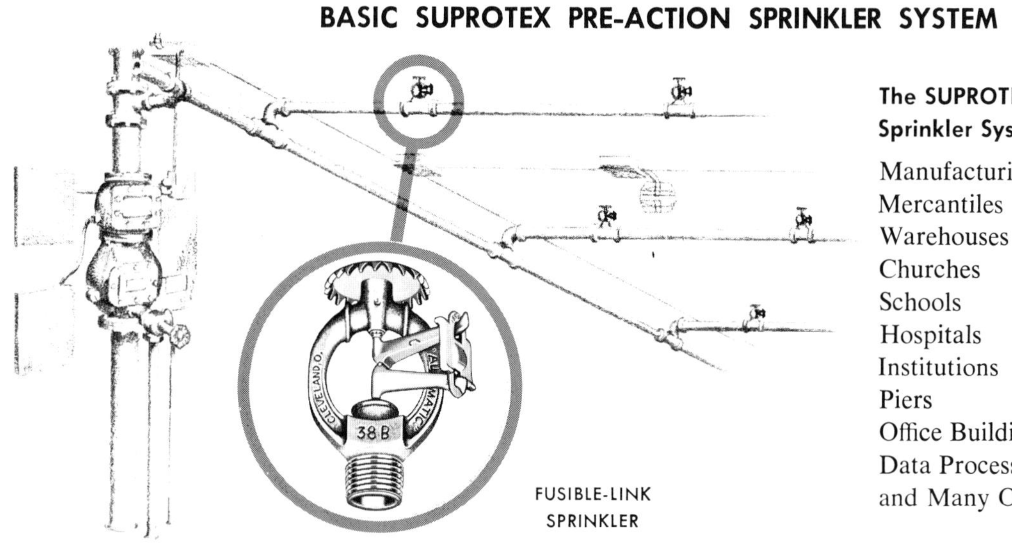

The SUPROTEX Pre-Action Sprinkler System for:

Manufacturing
Mercantiles
Warehouses
Churches
Schools
Hospitals
Institutions
Piers
Office Buildings
Data Processing Installations
and Many Other Applications

"Automatic" Sprinkler CORPORATION OF AMERICA • CLEVELAND, OHIO 44141

APPLICATIONS

OPERATION

The heat generated by a fire expands air in a Heat-Actuated Device increasing pressure. This pressure pushes against a diaphragm releasing a weight mechanism which trips the system control valve. The piping is immediately filled with water and a fire alarm is sounded by the water motor alarm gong. No water will discharge until a sprinkler is fused.

PRE-ACTION FEATURE

Pre-Action operation permits small fires to be extinguished manually, before any sprinklers are fused. High-value contents are protected against both fire and water damage.

A Pre-Action Sprinkler System stands ready to go into action if needed. The operation has placed water at the fusible-link sprinklers where it provides immediate spray in the event the premises are unattended and manual extinguishment is not possible.

If the sprinklers are not needed, the system is drained, the valve reset and the supervisory pressure restored.

SUPERVISED PRE-ACTION SPRINKLER SYSTEM

The SUPROTEX Pre-Action Sprinkler System provides constant supervision and notifies of possible damage. Both piping and air tubing are kept under a pressure of one and a half pounds per square inch, which is automatically maintained. The air tubing is supervised to assure operation; the piping is supervised to prevent accidental water damage.

PROTECTION DURING SERVICE OR REPAIRS

In case of damage to the sprinklers or piping, the sounding of the trouble horn warns that repairs must be made in order to avoid water discharge where not wanted when the system goes into action. The detection system can stay in service during this period.

If air tubing develops a slow leak, the Low Air Pressure Alarm will sound, indicating loss of pressure. The water supply does not have to be shut off while locating and repairing the leak. The valve can be tripped manually and operation depends on fusing of the sprinklers.

An abrupt loss of pressure in the thermal system, either from damage or an explosive fire, will trip the valve on loss of pressure because of the double-acting diaphragm. The piping fills with water, the fire alarm sounds, but there is no water discharge until a sprinkler head fuses.

UNSUPERVISED PRE-ACTION SPRINKLER SYSTEM

Where supervision is not required, the same system of early detection and alarm, plus Pre-Action operation of filling the system with water is available. This application is limited to small systems having 20 or less sprinklers.

The omission of automatic supervision means that, should damage occur to the air tubing or sprinkler piping, a trouble alarm would not sound. The Unsupervised Pre-Action Sprinkler System should be used only where material values and business interruption costs are not high.

COMBINED FIRE PROTECTION SYSTEMS

In areas that contain both ordinary and extra-hazard materials, a Combined System of Pre-Action (fusible-link sprinklers) and Deluge (open-head sprinklers) fire protection can be installed in unsupervised systems. Both are connected to the same valve.

4. PRE-ACTION FIRECYCLE SYSTEM (VIKING)

Action

1. Empty pipe branch lines become charged with water when an automatic detection system senses heat rise to a critical level (e.g. 135 or 140 degrees Fahrenheit).
2. Sprinkler heads open individually as heat reaches critical temperature for the head.
3. When fire is out or when heat subsides to 130–140 degrees at thermal sensor, system is prepared to close supply valve, which happens after a brief delay.
4. If fire rekindles and thermal sensors are heated again above 140 degrees F., water again flows through the branch pipe and any open heads.

Comments

1. Reduces possibility of accidental and unwanted discharge of water through damage to sprinkler pipes or heads.
2. Automatic closing of supply valve reduces automatically possibility of uncontrolled water damage.

The *Firecycle* system is described in the four pages following, reprinted from the Factory Mutual Record for March–April, 1967.

a sprinkler system that makes decisions

It has long been the dream of fire-protection engineers to design a sprinkler system that is completely automatic, that not only detects and attacks fire, but also automatically shuts off when the fire is out. Such a system is now a reality. The new decision-making system has been tested and is expected to be approved by Factory Mutual.

The development of the system over the past five years is an example of the traditional pattern of Factory Mutual cooperation with industry in providing new and better ways to conquer fire. The Viking Corporation of Hastings, Mich., with the Factory Mutual Research Corporation, conceived and designed the system. Fenwal, Inc. of Ashland, Mass. provided a special sensitive heat-detection switch and cable that will continue to operate reliably even after being subjected to extremely high temperatures.

In The Beginning

This fire-protection breakthrough is a far cry from the early days, a little over a century ago, when industry first became aware of the need for aggressive fire prevention and protection. The water bucket symbolizes the beginning of fire protection. Putting out fires was strictly manual in the early days, the most scientific approach being the formation of bucket brigades to transport the water more efficiently. When manufacturers in New England, tiring of disastrous fires, decided that something more effective was needed, the sprinkler system had its beginning. The first systems consisted of perforated piping that was installed in areas considered to be hazardous. These systems had piped water available, usually from private reservoirs containing a plentiful supply. However, perforated pipes wet down large areas and the human element was still a very dominant factor. The fire had to be discovered and the valve had to be opened. The man opening it had to quickly decide that the wetting of a large area of a plant was the only way to put the fire out; he had to decide when the fire was definitely out and when it was safe to close the valve, thus preventing further water damage. In other words, sprinkler systems were far from automatic.

The First Automatic Sprinkler Systems

The automatic sprinkler was invented in the latter part of the 19th century. The first sprinklers were crude and inefficient, but intensive activity from many sources gradually improved automatic sprinklers making them more reliable and effective over the years until today's standards were attained.

Yet today's systems are, despite their name, only semiautomatic. True, the sprinklers detect fire, open automatically and often extinguish fires without human assistance. One automatic feature, however, is still lacking; the decision still has to be made to manually close the valve at the right time and to reopen it if the fire rekindles. Some of this century's biggest fires developed because the valve was closed too soon or not reopened when the fire rekindled. In a recent decade, the damage resulting from 14 such fires came to 32,000,000 dollars; one loss alone amounted to 11,000,000 dollars. A well-trained emergency organization does not let this happen. However, the possibility of human error always exists and can catch unwary plants off guard. The new sprinkler system should forestall such occurrences.

Fig. 1. This 1878 sprinkler proves that from the beginning fully automatic sprinkler systems were the goal of inventors. The sprinkler contained a liquid that expanded with heat, forcing the sides apart and allowing water to flow. When the heat of the fire died down, the sprinkler was supposed to close again. Unfortunately, the cold water coming out closed the sprinkler before the fire was out.

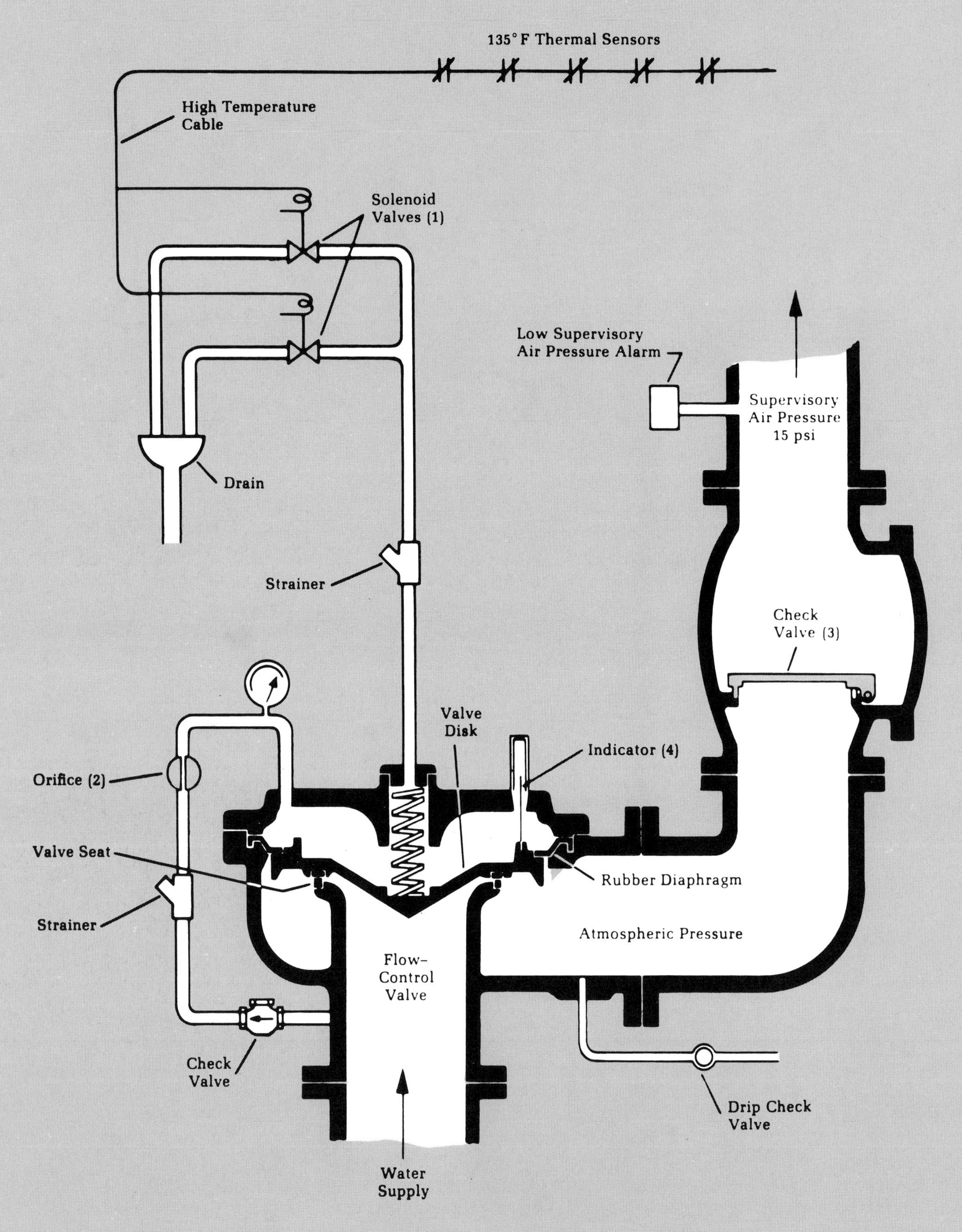
135° F Thermal Sensors
High Temperature Cable
Solenoid Valves (1)
Low Supervisory Air Pressure Alarm
Supervisory Air Pressure 15 psi
Drain
Strainer
Check Valve (3)
Valve Disk
Indicator (4)
Orifice (2)
Valve Seat
Rubber Diaphragm
Strainer
Atmospheric Pressure
Flow-Control Valve
Check Valve
Drip Check Valve
Water Supply

Fig. 2.

The Firecycle System — How It Operates

The recently tested Firecycle system offers to industry the first fully automatic sprinkler system. The heart of the system is a hydraulically operated valve. It serves all sprinklers fed by a single riser and operates much like a dry-pipe valve except that water pressure at the valve itself, rather than air in the piping above the valve, keeps it in the closed position. By its construction the 6-in. valve has a differential of 3 to 1. For example, with water pressure of 90 pounds above and below the valve, the valve will not open until the pressure above it has been reduced by two-thirds

Fig. 2. A schematic sketch showing the fundamental operating principle of the Viking Firecycle Sprinkler System. The flow-control valve is in the closed position. The two energized solenoid valves (1) which are normally closed, open when thermal sensors reach about 140° and open the electrical circuit. Opening of these valves drains the water pressure from above the flow-control valve faster than the restricted orifice (2) can replenish it. Water pressure below the valve then pushes it upward from its seat and sends water into the sprinkler system where it will be immediately available when sprinklers fuse. The purpose of the check valve (3) is to hold supervisory air on the system; it opens immediately when water enters the system.

When the fire has been extinguished or nearly so, the thermal sensors cool and make contact again at about 130-140°F. As a safety factor, a timer keeps the solenoid valves open for a predetermined time before they are energized and again close. Pressure then builds up through the restricted orifice (2) and in about 60 seconds the water pressure above the flow-control valve is equal to that below the valve and it drops back into the closed position. Water remains in the sprinkler system above the check valve ready for action. If the sensors again are heated until the contacts separate, the solenoid valves open, sprinklers operate, and the cycle automatically repeats. Two solenoid valves are used in parallel; the second increases reliability by four times. An indicator (4) shows at a glance whether the flow-control valve is open or closed.

Fig. 3. Firecycle flow-control valve and fittings.

to about 30 pounds. Sprinkler piping above the valve is empty except that low air pressure is maintained for supervisory purposes. When a fire occurs, thermal sensors break electrical contacts at about 140°F and cause two normally closed solenoid valves to open. These valves relieve the pressure above the flow-control valve which then opens and the system fills with water well before the automatic sprinklers reach operating temperature.

When the fire is under definite control and temperatures at the sensors decrease to under 140°F, their contacts close again. After the last contacts close, an automatic timer allows the system to continue to operate for a predetermined time, usually five minutes, as a safety factor. Then the solenoid valves close again, pressure begins to build up above the flow-control valve and shuts it in about 60 seconds. If the fire rekindles after the flow control has shut, the sensors immediately reopen the valve. Since the sensors are more sensitive and operate at lower temperatures than sprinklers it is unlikely that additional heads will operate if the fire rekindles. After the fire the flow-control valve shuts automatically. Replacement of sprinklers does not interrupt automatic protection. They may be replaced soon after the fire without shutting off the water-supply valve. The system remains operative throughout.

Other Features

- *The system is designed to "fail safe."* If electrical power fails, the system automatically goes on battery power and remains fully automatic. Should the battery power run out before electrical power is reinstated, the solenoid valves open automatically and the system becomes an automatic wet pipe system and a water-flow alarm sounds. The battery will provide power for at least four days. If the electrical system must be disconnected for electrical repair, a two-position valve is moved by hand into "standby" making the system an automatic dry-pipe system. Under the two latter conditions the flow-control valve will not shut automatically, but it will open automatically and protection is maintained.
- *Valve closures at a bare minimum.* There is no need to close the post indicator, or other manual valve for the system, except for maintenance of the flow-control valve or piping leading to it. Changes in sprinkler piping may be made without closing any valves. Thus, in addition to preventing prematurely closed valves during fires which cause some of our most damaging fires, the system should help reduce the number of fires which result from fire striking and finding the valve shut.
- *Chances of accidental water damage reduced.* Accidental damage to the sprinkler system, such as might be caused by careless operation of a lift truck, or by freezing of water in low points will give an alarm to indicate something is wrong, but there will be no water damage. Low air pressure, above an approved check valve, supervises the system and through a low-pressure alarm switch indicates trouble on drop of pressure.
- *Water is used efficiently.* Because the system, unlike humans, can sense conditions in a fire area despite heat and smoke, it can decide for itself when it is safe to shut off and water is conserved.

What Are The Costs?

A recycling sprinkler system with a flow-control valve is expected to cost somewhat more to install than the present conventional system. The additional cost will have to be weighed against the value of a greater probability that a building's fire-protection system cannot be nullified by an error in judgment. There are other advantages. For example, in hazardous plants or areas where materials or equipment are particularly prone to water damage, the additional cost might be written off alone by the decrease in damage caused by water flowing during the period when it is not certain that the fire is out.

Inspection and Maintenance

Reliability was the prime consideration when components for the system were chosen and they are expected to give many years of trouble-free service. Periodic inspection and maintenance are needed at specified intervals to determine, for example, that orifices are free and that all moving parts are operative. The system is easy to test by heating one of the thermal sensors, or by pushing a "thermal test" button, which opens the electrical circuit and causes the solenoid valves to open. This causes water to fill the system and operate the alarms. The valve then resets itself and the system can be drained again and returned to service.

A Decisive Protection Choice

The new sprinkler system which will be in production this year provides another new dimension in reliable fire protection. The chances of human error are reduced, improperly closed valves diminish as a factor in disastrous fires and water is used efficiently.

5. AQUAMATIC SPRINKLERS (ON-OFF) BY GRINNELL

Action

1. Sprinkler heads are on pipes charged with water and designed to discharge at 165 degrees Fahrenheit.
2. Heads are opened individually by action of a bi-metallic snap disc.
3. Heads close again individually when fire is under control and temperature drops to around 100 degrees F. No adjustment is needed; the head is ready for further cycling if fire rekindles.

Comments

1. Newest of systems.
2. Individual operation reduces possibility of excess water.
3. Original (1973) head was redesigned to achieve higher performance levels in 1975 model.
4. Underwriters' Laboratories (UL) listing for use on preaction (empty pipe) systems has been recently applied for and the head has passed the required testing procedures.

The *Aquamatic* Sprinkler is described in the bulletins on the next four pages.

GRINNELL FIRE PROTECTION SYSTEMS COMPANY, INC.

Aquamatic Sprinklers (On-Off)

Style: Pendent

Pendent

The first fully approved automatic On-Off sprinkler for continuous, dependable fail-safe fire protection. It's the only sprinkler available that resets itself automatically after it extinguishes a fire. It's set to go time after time without replacement or adjustment. There's no need to turn off the main valve for inspection after a fire, which eliminates downtime and still gives 100% fire protection all the time. The system is always up and ready to go.

Applications

The Grinnell Aquamatic On-Off Sprinkler is designed for wet pipe systems. It is ideal for areas containing high value inventories or materials highly sensitive to water; and in areas where there is a risk of flash fires, repeat fires, or where the water supply is limited. Typical installations include: warehouses, plant spray booths, computer rooms, record storage facilities, retail outlets, high rise apartments and office buildings, hospitals and nursing homes. The Aquamatic Pendent sprinkler is used when piping is close to the ceiling or concealed in the ceiling.

Typical computer room installation

Description

The Grinnell Aquamatic Sprinkler System consists of fully automatic On-Off sprinklers mounted at strategic intervals on a network of overhead piping. Water is fed through the piping system which is graded in size to assure adequate water pressure to the protected area. Each head is fully automatic, which means water is discharged only when it is needed and only as long as it is needed. When the heat from the fire is removed by the discharged water, the sprinkler automatically turns off and is ready to reactivate with no readjustment or replacement. The sprinklers sequentially turn themselves on and off as needed. There is no need to manually shut off the main control valve to arrest the water flow. This is especially important where there is limited water supply or a reservoir.

Features

- No replacement or adjustment.
- Efficient use of water.
- Helps eliminate main valve closing.
- Totally automatic.
- Fast response.
- Interchangeable with existing systems.
- UL listed; FM approved.

New Issue: Aug. 1975
Supersedes: Dec. 1973

Component Parts 1. Inlet 2. Body. 3. Piston 4. B/AS Spring 5. Pilot Valve 6. Yoke

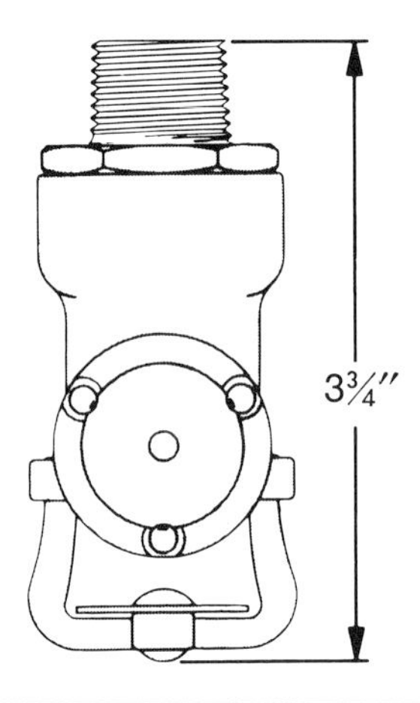

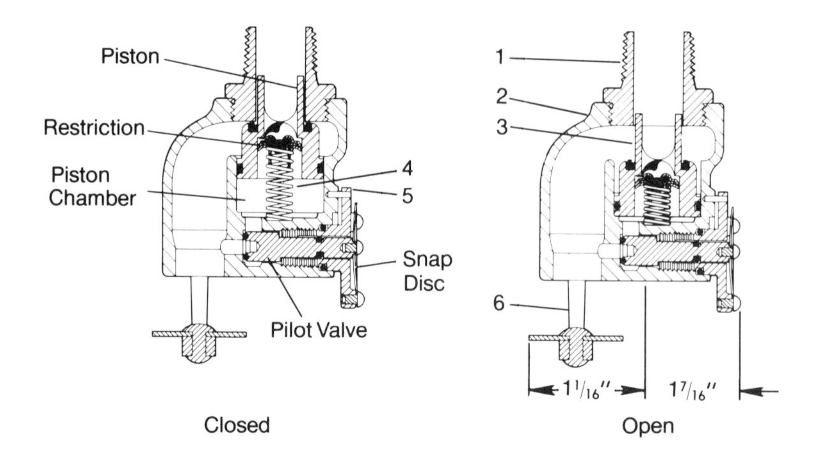

Operation

When the bi-metallic disc is exposed to heat in excess of 165°, it opens to allow a small amount of water to pass through the pilot orifice. This action releases pressure on the piston assembly. Water is actually released faster than it can be replenished through the restrictor, which forces the piston down allowing water to discharge onto the fire through the main port. When the fire is controlled and the heat subsides to approximately 100°, the disc snaps in again and closes the pilot orifice. At this point, water pressure builds up in the pressure chamber and forces the piston assembly closed. The flow of water is arrested and the sprinkler resets itself automatically. There is no further need to readjust or replace the sprinkler heads.

Tests

An exhaustive series of tests were conducted by FM and UL to prove Aquamatic's performance under the severest of conditions.

Operational

Aquamatic is the first and only On-Off sprinkler to pass all the standard sprinkler tests for Factory Mutual and Underwriters Laboratories. These included operational tests for temperature, for calibration, for vibration, for corrosion, for pressure, for leakage, and for water distribution. Aquamatic passed them all. With flying colors. Then it was subjected to another series of tests that other sprinklers never go through. The purpose was to prove its performance ability under the severest conditions.

Cycling

A solution of rust and scale was introduced into the water supply of a system using Aquamatic sprinklers. The sprinklers were then subjected to heat and allowed to operate through a thousand cycles. They turned on and off without fail every single time. There were no malfunctions. No defective parts. There was no clogging, either.

Fire

FM and UL put Aquamatic through its paces in actual full-scale fire tests, too. The sprinklers were installed with 10′ by 10′ spacing over a pallet storage area. The pallets held 90 corrugated cartons encompassing an area 12′ high, 12′ wide, and 40′ long with 6″ of free space between stacks.

The fire was started at floor level between the first and second cartons.

During the test, water flow through the system was maintained at 30 gpm per sprinkler head and the temperature never exceeded 495° at any point over the whole area. When the fire was officially out, the total fuel consumed was well below the FM and UL requirements. All the Aquamatic sprinklers had performed flawlessly during the fire, reset themselves when it was out, and were ready to go again.

Specifications

Temperature Rating: 165°, 212°
Deflector Type: SSP-3 (Pendent)
Pipe Thread Connection: ½″ NPT
Orifice Size: ½″
Finish: Plain Brass, Chrome
To Specify: Grinnell On-Off Sprinkler Model F920—Pendent, (temp. rating), (finish) and quantity.

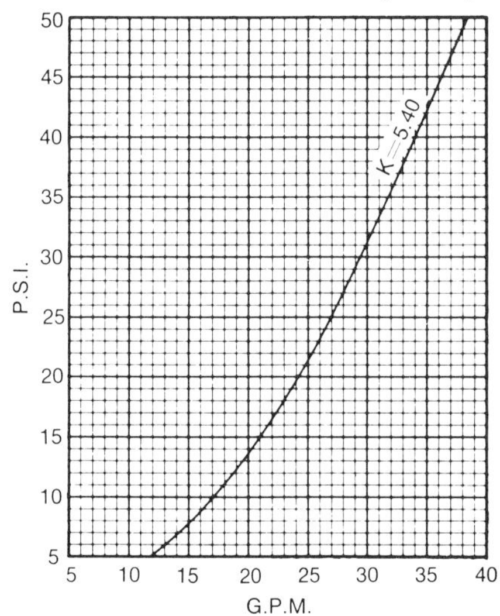

Discharge Curve Graph

GRINNELL

GRINNELL FIRE PROTECTION SYSTEMS COMPANY, INC.

Aquamatic Sprinklers (On-Off)

Style: Recessed

Recessed

The first approved automatic On-Off recessed sprinkler that combines continuous, dependable fail-safe fire protection with pleasing appearance. It's the only sprinkler available that resets itself automatically after it extinguishes a fire. It's set to go time after time without replacement or adjustment. There's no need to turn off the main valve for inspection after a fire, which eliminates downtime and still gives 100% fire protection all the time. The system is always up and ready to go.

Applications

The Grinnell Aquamatic On-Off sprinkler is designed for wet pipe systems. Because it's a recessed sprinkler, it's designed for protecting areas where architectural appearance must not be compromised, and is ideal for use with high value inventories, or materials highly sensitive to water and in areas where there is a risk of flash fires, repeat fires, or where the water supply is limited. The recessed Aquamatic is typically used in applications that include restaurants, hotels, banks, lobbies, churches, school halls and corridors, auditoriums, reception areas, shopping malls, computer rooms, and public areas.

Typical Aquamatic Recessed Installation

Description

The Grinnell Aquamatic Sprinkler System consists of fully automatic recessed On-Off sprinklers mounted at strategic intervals on a network of overhead piping. Water is fed through the piping system which is graded in size to assure adequate water pressure to the protected area. Each sprinkler is fully automatic, which means water is discharged only when it is needed and only as long as it is needed. When the heat from the fire is removed by the discharged water, the sprinkler automatically turns off and is ready to reactivate with no readjustment or replacement. The sprinklers sequentially turn themselves on and off as needed. There is no need to manually shut off the main control valve to arrest the water flow. This is especially important where there is limited water supply or a reservoir.

Features

- Appearance.
- No replacement or adjustment.
- Efficient use of water.
- Helps eliminate main valve closing.
- Totally automatic.
- Fast response.
- Interchangeable with existing system.
- UL listed.

New Issue: Aug. 1975
Supersedes: Dec. 1973

Component Parts:

1. Body
2. Adjusting Collar
3. Orifice Cover
4. Flat Head Screw
5. Cover Plate
6. Aquamatic Sprinkler

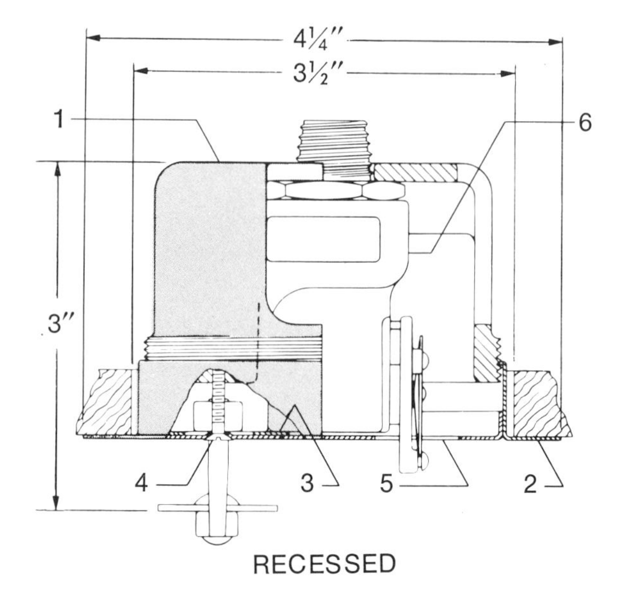

RECESSED

Component Parts:

1. Inlet
2. Body
3. Piston
4. Spring
5. Pilot Valve Assembly
6. Yoke & Deflector Assembly

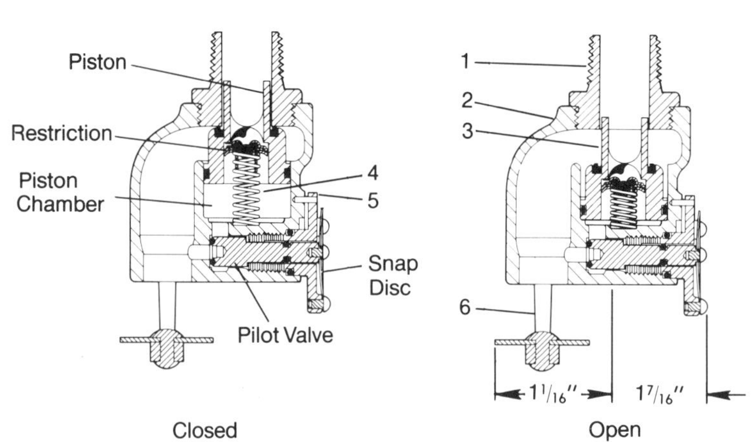

Closed Open

Specifications

Temperature Rating:	165°, 212°
Deflector Type:	SSP-3
Pipe Thread Connection:	½″ NPT
Orifice Size:	½″
Finishes:	Chrome
To Specify:	Aquamatic Recessed Sprinkler Model F 922, and quantity.

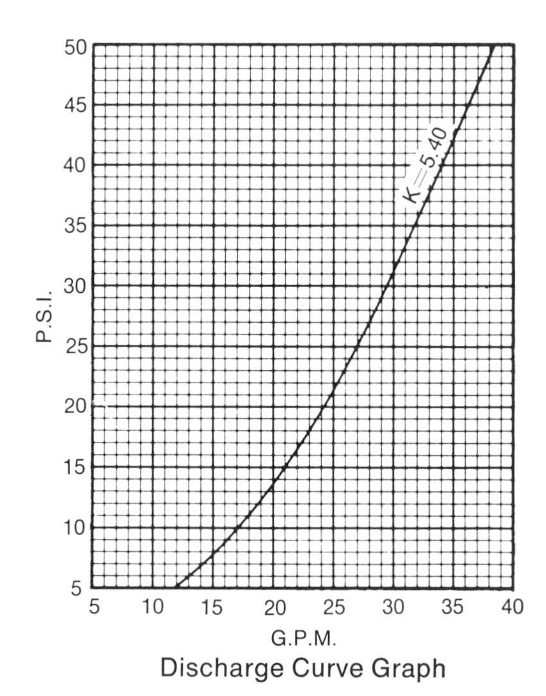

Discharge Curve Graph

6. HALON 1301—TOTAL FLOODING SYSTEM

Action

1. A halogenated hydrocarbon liquefied vapor ($CBrF_3$, bromotrifluoromethane) is held in steel containers under pressure and connected to piping with open nozzles.
2. Various means of discharging the system are used, including automatic detection of heat, flame, smoke, or by other means.
3. A fire is extinguished immediately by a chemical process of interference with the fire's chain reaction.

Comments

1. No moisture or other residue is deposited; the agent is considered particularly appropriate for rare book rooms.
2. The 5% or 6% Halon 1301 atmosphere sufficient to kill most fires is harmless to people in exposures as long as five minutes.

Halon 1301 systems and hardware are described in the following four pages, reprinted from the January/February 1971 *Dupont Magazine*. There are a number of manufacturers of systems for application of Halon 1301 whose products are listed in publications of the testing laboratories, Underwriters' Laboratories and Factory Mutuals.

New Fire Protection for High Value Areas

Automatic total-flooding systems use Du Pont "Freon" 1301 fire extinguishant for efficient control

BY GEORGE NEILSON

Behind the 107-year-old walls of the four-story row house on Philadelphia's Delancey Place is one of the world's finest private collections of rare books, manuscripts, and historic documents. It is invaluable, irreplaceable, and highly combustible. For 16 years, the trustees of the Rosenbach Foundation Museum lived uneasily with the haunting possibility of fire.

When the younger of the two Rosenbach brothers received his Ph.D. in English literature from the University of Pennsylvania in 1901, he acquired the name he would be called for the rest of his life: Doctor. He also acquired an erudition that made him the world's most successful dealer in rare books and manuscripts. Between 1900 and 1950, he engaged in the purchase and sale of more than $100 million worth of books and manuscripts, among them some of history's most important written documents.

Philip, the older brother—a dealer in silver and furniture—sometimes chided Doctor for becoming overly attached to a particular acquisition. After all, he would point out, business *was* business; the documents were items of trade. On such occasions, Doctor would reply in kind: hadn't Philip kept for himself a handsome 18th century candelabrum? They would go for days without speaking; Philip in his rooms on the second floor, Doctor in his on the third. In spite of one another, however, each built up a priceless collection.

Reprinted from
Jan./Feb. 1971
DU PONT
magazine

In 1954, the Rosenbach Foundation was created. The brothers had willed to it their home, the two lifetimes of acquisitions it housed, and their money to maintain the museum and to purchase new treasures for it. Today, the contents of the quiet townhouse reflect centuries of man's insatiable curiosity, his expanding intellectualism, and his ability to speak not only for himself but for his fellow man.

In the house, where the atmosphere is controlled to a precise degree of temperature and humidity, are rare paintings and drawings—many of them originally book illustrations—by Fragonard, Turner, Blake (an illustration for the Book of Revelations), George Bellows and others. Glass display cases contain original manuscripts by Chaucer (*The Canterbury Tales*), Dickens (*The Pickwick Papers*), and James Joyce (*Ulysses*). Among the important books are the *Bay Psalm Book,* printed in 1640, the first book printed in what is now the United States.

There are letters, often touchingly human, often of major historic interest; e.g., a farewell to her retinue by Mary Queen of Scots on her imprisonment; a discussion of family affairs with his half-brother Lawrence by 17-year-old George Washington in 1749, the earliest known Washington letter in existence; a letter from Hernando Cortés to the Court of Spain, asking permission to explore the land that is now California. In General Grant's handwriting is the text of his telegram to Secretary of War Stanton, informing him of Lee's surrender.

"There is no way to place a value on the

continued

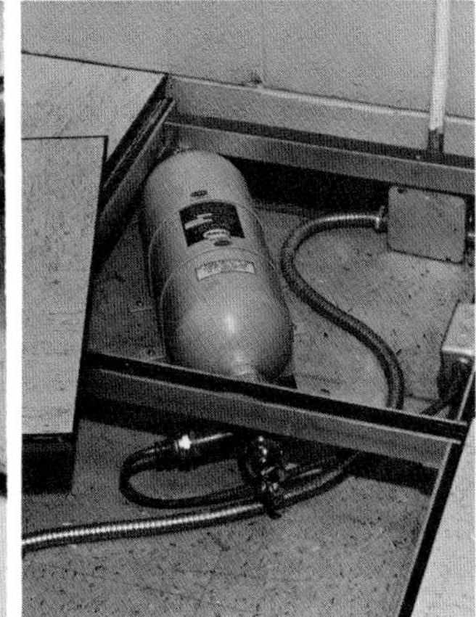
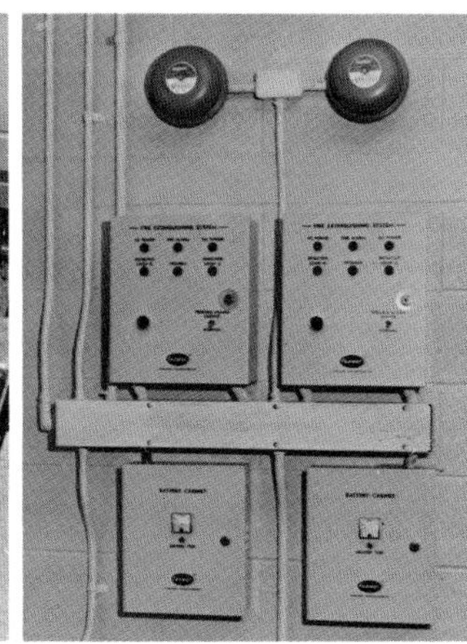

In computer rooms, "Freon" 1301 can be stored under floor, activated by sensing device, and released through ceiling nozzles. Extinguishant leaves no residue that might harm equipment.

contents of the museum," says director Clive E. Driver. "And, of course, we worry about the possibility of fire. The trustees have often discussed moving the collection to a modern, fireproof building, but the present home—with its dignity, elegance, and charm—seems so right. We decided to stay here and install the best fire protection we could get."

There had never been a fire-extinguish ing system in the building—only an alarm system to summon firefighters. "Water sprinklers were out of the question," Driver says. "Water would have caused as much damage as the flames, perhaps more. What we needed here, with so much at stake, was a completely new kind of system."

Last summer, such a system—designed to flood an enclosed area with a highly effective vapor—was installed by James Castle, Inc., Philadelphia. Strategically and inconspicuously placed brass nozzles, connected by pipes to a central source of extinguishing agent, stand ready to release the vapor as soon as sensing devices detect fire. "Our biggest problem," says Castle's president, Charles Ford, "was to get the piping through the old building without destroying its beauty or function."

The effectiveness of the system, Ford points out, depends on "Freon" 1301 fire extinguishant, a liquefied, compressed gas. Chemically, the product is bromotrifluoromethane, known in the trade as Halon 1301. Unlike water, which puts out fires by cooling, or carbon dioxide, which smothers flames, "Freon" 1301 does its work chemically. It actually interferes with the combustion cycle, stopping it immediately. This quick action can prevent a fire from getting out of control.

After early success in protecting racing cars, automatic systems using 1301 were slow in finding widespread acceptance. "One reason," says Du Pont product manager David Fariss, "was that the National Fire Protection Association did not have a standard for the use of the product until May 1970. Now it is permitted for the three major classes of fires: Class A, cellulosic materials; Class B, flammable liquids; and Class C, electrical. Properly used, it is the safest gaseous extinguishing agent available."

Underwriters' Laboratories classifies "Freon" 1301 in Group Six, the category reserved for the least toxic of materials tested. Systems using it can be discharged even while people are in the vicinity.

Another obstacle to rapid growth of the use of "Freon" 1301, Fariss points out, was the lack of hardware approved by Factory Mutual, Underwriters' Laboratories and similar organizations . . . the pipes for carrying the gas, the nozzles, and the devices which detect the presence of fire and release the extinguishing agent. "Within the past year," he says, "a number of leading manufacturers of fire protection systems have developed approved, efficient equipment. Installations are increasing at a steady rate in 'high value' areas. These are places where the contents of a room or building are extremely valuable or irreplaceable, and where the use of water or other extinguishing agents could cause extensive damage."

Industry leaders predict a steady but cautious growth for the new system. "In our opinion," says George Grabowski, manager, Protection System Division, Fenwal, Inc., Ashland, Mass., "1301 is the fire extinguishing agent of today and tomorrow. What we in the industry must do is to see that it is used properly, that it does not become a gimmick indiscriminately used. Systems must be carefully designed for the areas they are to protect; hardware should be especially suited to 1301 systems."

Fenwal, with an extensive background in developing and manufacturing fire and explosion suppression systems for industry, first used "Freon" 1301 in its portable "Firepac" units—widely used during the construction and maintenance of commercial and military aircraft. Following this success, Fenwal moved to fixed systems using 1301. One of these protects a vault containing more than $5 billion in negotiable securities; another is installed in a Saskatchewan mine to protect essential equipment. Other Fenwal systems protect computer rooms, fuel storage tanks, and communications and utility installations.

In most metropolitan areas, utility companies maintain "peaking stations" with emergency generators that can cut in, automatically, to handle sudden demands for more electric power. That often happens when there is a sudden increase in the use of air conditioners. Many times these stations are unattended. A fire in one of them

Portable units using "Freon" 1301 extinguishant protect valuable cargo in truck shipment.

could interrupt electric service, throw a city into darkness, leave elevators suspended between floors, bring industrial operations to a halt. Automatic Fenwal systems using 1301 are installed in many of these stations to extinguish such fires as soon as they break out.

Similarly left unattended are many telephone switching stations handling millions of dialed calls each day. Again, fire could disrupt operations; residue from extinguishing agents could delay the return of service. And again a Fenwal system provides silent, constant protection. In both kinds of installations, the moment the sensing devices detect fire and release the gas, they sound an alarm at headquarters. Crews are then dispatched to investigate the situation and get the equipment operating again so service can be resumed.

Computer installations seem a particularly appropriate application for 1301, according to Robert Sunstrom, assistant to the director of planning, research, and development, for the Ansul Company, Marinette, Wis. "There are some 80,000 computer installations in the U.S.," he estimates, "and of these probably less than 1,000 have any kind of fire protection system. One reason, of course, is that until now there hasn't been a system that was truly compatible with computers. Dry chemicals, once considered the most effective agent for electrical fires, leave a residue of small particles which can damage sensitive equipment. Carbon dioxide doesn't make a mess, but it can produce condensation which ruins equipment."

Ansul, one of the nation's leaders in engineered fire protection, began experimenting with the halogenated compound agents more than 15 years ago, and introduced its "clean agent" system utilizing 1301 last July. "We waited until we could offer a complete protection package," Sunstrom says. "We have it now: hazard analysis, design, piping, installation, and the option of nearly any type of detection device." In addition to computer centers, Sunstrom expects the Ansul system to find broad application in radio and television stations, in record storage rooms — "wherever residue problems from traditional extinguishing agents have restricted the installation of adequate fire protection."

Neil Hoogmoed, product manager for Halon Division, Walter Kidde and Co., Inc., Belleville, N.J., points out that while "Freon" 1301 is a new extinguishing agent to many in industry, "it is an old friend to many pilots. Our Aerospace Systems Division has designed these systems for years."

The low toxicity characteristic of 1301 is probably the most important factor in its selection where human safety is involved. According to Hoogmoed, it was the chief consideration in installations on Alaska's North Slope. "Up there," he says, "oil drilling is a lot different than it is in most locations because of the severe climate conditions. It calls for new and more extreme safety measures than ever before used in the oil industry."

In large fields, Hoogmoed explains, it is common practice for oil to be pumped from a number of wells into a central "gathering station" where oil is separated from gas. In most places, this is done in the open air where fumes are quickly dispersed. In Alaska, where temperatures are often 50 below zero and savage snowstorms are frequent, the separation is done inside a series of inter-connecting buildings. Because of the high ratio of gas to oil in the North Slope fields, concentration of fumes in the gathering stations can rise to dangerous levels. To protect personnel and equipment from explosion and fire, 1301 systems were specified for some installations at Prudhoe Bay. Walter Kidde's Halon Division supplied them.

In a gathering station, a sensing device called a Gas Analyzer samples the air to determine the percentage of gas present. If the concentration reaches 25 per cent of the lower explosion limit, an alarm sounds, vents open, and exhaust fans are turned on. If the concentration builds to 75 per cent, the Analyzer closes vents, turns off fans, and releases "Freon" 1301 within fractions of a second.

Instrument room installations offer protection against fire damage and resulting downtime.

There are other reasons for the use of "Freon" 1301 on the North Slope. It can penetrate every crevice in an enclosed area to make the atmosphere inert; there are no residual particles to foul up machinery. The system is about one-fourth the size of a comparable carbon dioxide system and can be installed, generally, at lower cost.

Other Walter Kidde, Halon Division, installations are in film storage vaults, gas compressor stations, computer rooms, and insurance policy storage rooms. "Wherever risk and value are high," Hoogmoed says, "a 1301 system is almost essential. Recently, we installed a system in a credit card center for a large oil company. Every day, customer charge slips amounting to over $100 million come into its vaults. The slips are the only records the company has of the transactions. Imagine the loss if they went up in flames!"

Leaders in the fire protection industry generally agree that 1301 has had a tremendous impact in a relatively short time. "There is little doubt," says Fenwal's Grabowski, "that it will change the face of fire protection in this country and abroad."

There is also agreement that for the immediate future, at least, the uses of 1301 will be restricted largely to "high value" applications. "Let's face it," says Castle's Charles Ford, who has been affiliated with fire protection equipment for more than 40 years, "for many applications, the water sprinkler system is adequate, and much less expensive. But where you have special hazards—where people are willing to pay the price of proper protection—that's where 1301 will find the greatest acceptance."

Today, fire protection systems using "Freon" 1301 cost about $500 per thousand cubic feet of space. There is hope that in time costs will decrease. "But how much is too much," asks Du Pont's Fariss, "when the preservation of a Dickens' manuscript, the continued functioning of electronic equipment, or—above all—a human life is at stake?" ■

7. AUTOMATIC DETECTION SYSTEMS

Action

1. Sensors are placed usually at ceiling height and spaced according to specifications of Underwriters' Laboratories.
2. Sensors detect fire by recognizing:
   (a) Products of combustion (invisible and visible);
   (b) Visible products of combustion (smoke);
   (c) Rapid rise of heat (rate of rise detector);
   (d) Fixed temperature attained.
3. A detection system can operate various systems (close doors, shut down motors, etc.) but almost invariably is used to send a warning message to a fire department or central station.

Comments

1. Engineering judgment is needed for decisions on the system appropriate to the risk.
2. These systems warn of smoke or fire—they do not put a fire out. Effective control of fire depends on the fire fighting process, whether through a mechanical system or a fire department. The warning process is only an auxiliary system.

Typical fire detection hardware and systems are shown on the following two pages from a publication of Pyrotronics, Inc., one of several manufacturers.

HOW SMALL FIRES BECOME BIG ONES

DELAYED DETECTION

Every big fire started as a small fire . . . that was not detected early enough to nip it in the bud. In fact, 89% of all large fire losses ($250,000 or more) became large because of a delay in alarm.

Most large loss fires occur in schools, churches, stores, plants, and warehouses when they are not in operation, such as, after working hours, during holidays or on weekends. Incipient conditions are "locked up" at closing time, giving the fire plenty of time to make headway. Even during business hours, fires in unoccupied or concealed areas often go "undetected" until it is too late.

DELAYED ALARM

When the fire is detected, employees often attempt to fight the blaze and neglect to turn in an alarm—wasting precious time. Some people panic, and are unable to take any effective action at all. And even when an alarm is turned in quickly, it may be telephoned incoherently or incompletely, putting an additional handicap on the fire department. The result is critical time lost in summoning competent fire fighters.

EARLY DETECTION AND ALARM

Fire department officials are unanimous in their appeal for a system of fire detection that will give an alarm in the earliest stages of combustion.

Many times there are only a few minutes between the beginning of combustion and the development of a truly destructive fire. These precious minutes are all that is required to counteract the headway of the flames. Even more important is the requirement for the safe evacuation of the building and the probability that portable equipment can satisfactorily extinguish or control the fire before the arrival of firemen.

ALARM INITIATING DEVICES

FIRE DETECTOR BASE

The detector base is the key to the flexibility of the Pyr-A-Larm system. Constructed of durable, molded plastic, it is designed to accept all four types of Pyr-A-Larm plug-in detectors. All connections are screw type, eliminating soldered connections. The base also contains a neon pluse lamp which identifies the detector initiating the alarm. It is 4½" in diameter and mounts on a standard 4" octagon box.

IONIZATION FIRE DETECTORS

Operates on a patented ionization principle that reacts to combustion gases, the first stage of fire. Does not require heat, flame or visible smoke to operate. Available in two models for use with either high or low movement of air. Recommended for protection of general areas or specific equipment.

FLAME FIRE DETECTOR

Instantly senses infrared radiation emanating from flames. Intended for areas of fast developing fires where ignition is almost instantaneous, high ceiling areas or high air movement areas. To prevent false alarms, unit is sensitive only to flame flickering sustained 3, 10 or 30 seconds, depending on detector selected.

PHOTOELECTRIC FIRE DETECTORS

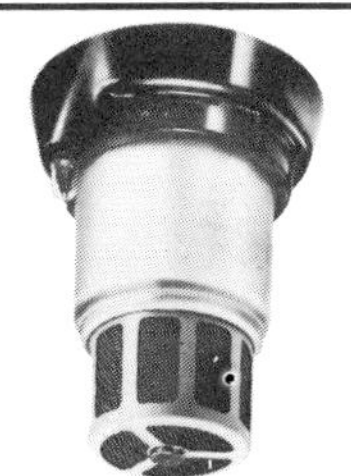

Responds directly to visible smoke. Intended for areas where it is not practical to use ionization detectors due to normal high ambient level of combustion gases or where material protected will produce heavy smoke. Available in three sensitivity ranges of 0.4/ft., 1%/ft. and 3%/ft. obscuration.

PLUG-IN THERMAL FIRE DETECTORS

Eight thermal detectors are available in four temperature ratings—fixed or combination, fixed temperature and rate-of-rise,—either non-restoring or self-restoring. These thermal detectors fit the standard Pyr-A-Larm base and are instantly interchangeable with ionization, flame or smoke detectors.

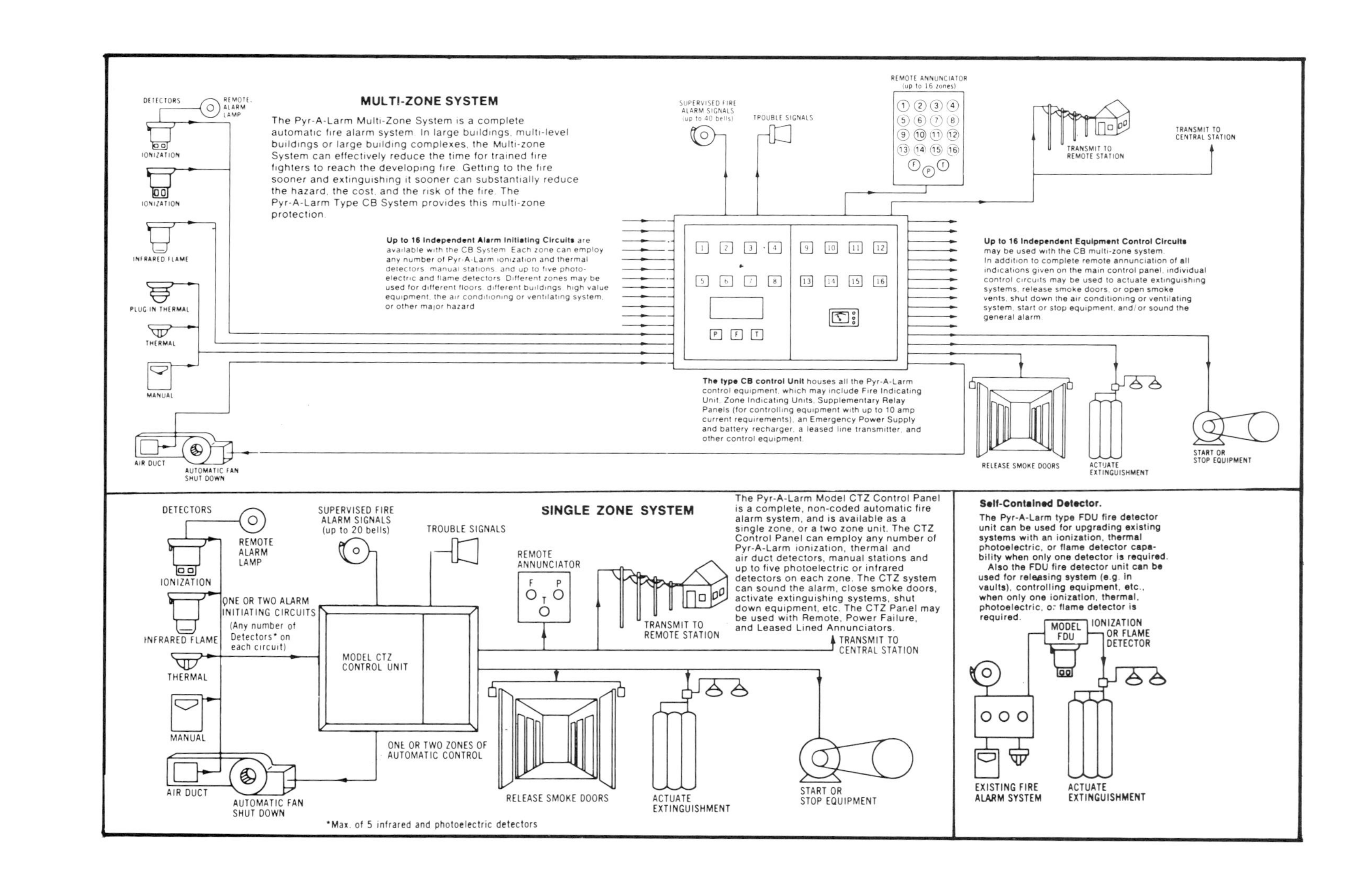
MULTI-ZONE SYSTEM
The Pyr-A-Larm Multi-Zone System is a complete automatic fire alarm system. In large buildings, multi-level buildings or large building complexes, the Multi-zone System can effectively reduce the time for trained fire fighters to reach the developing fire. Getting to the fire sooner and extinguishing it sooner can substantially reduce the hazard, the cost, and the risk of the fire. The Pyr-A-Larm Type CB System provides this multi-zone protection.
DETECTORS
REMOTE ALARM LAMP
IONIZATION
IONIZATION
INFRARED FLAME
PLUG IN THERMAL
THERMAL
MANUAL
AIR DUCT
AUTOMATIC FAN SHUT DOWN
Up to 16 Independent Alarm Initiating Circuits are available with the CB System. Each zone can employ any number of Pyr-A-Larm ionization and thermal detectors, manual stations, and up to five photoelectric and flame detectors. Different zones may be used for different floors, different buildings, high value equipment, the air conditioning or ventilating system, or other major hazard.
SUPERVISED FIRE ALARM SIGNALS (up to 40 bells)
TROUBLE SIGNALS
REMOTE ANNUNCIATOR (up to 16 zones)
TRANSMIT TO REMOTE STATION
TRANSMIT TO CENTRAL STATION
Up to 16 Independent Equipment Control Circuits may be used with the CB multi-zone system. In addition to complete remote annunciation of all indications given on the main control panel, individual control circuits may be used to actuate extinguishing systems, release smoke doors, or open smoke vents, shut down the air conditioning or ventilating system, start or stop equipment, and/or sound the general alarm.
The type CB control Unit houses all the Pyr-A-Larm control equipment, which may include Fire Indicating Unit, Zone Indicating Units, Supplementary Relay Panels (for controlling equipment with up to 10 amp current requirements), an Emergency Power Supply and battery recharger, a leased line transmitter, and other control equipment.
RELEASE SMOKE DOORS
ACTUATE EXTINGUISHMENT
START OR STOP EQUIPMENT
SINGLE ZONE SYSTEM
DETECTORS
REMOTE ALARM LAMP
IONIZATION
ONE OR TWO ALARM INITIATING CIRCUITS
(Any number of Detectors* on each circuit)
INFRARED FLAME
THERMAL
MANUAL
AIR DUCT
AUTOMATIC FAN SHUT DOWN
SUPERVISED FIRE ALARM SIGNALS (up to 20 bells)
TROUBLE SIGNALS
MODEL CTZ CONTROL UNIT
ONE OR TWO ZONES OF AUTOMATIC CONTROL
REMOTE ANNUNCIATOR
TRANSMIT TO REMOTE STATION
TRANSMIT TO CENTRAL STATION
The Pyr-A-Larm Model CTZ Control Panel is a complete, non-coded automatic fire alarm system, and is available as a single zone, or a two zone unit. The CTZ Control Panel can employ any number of Pyr-A-Larm ionization, thermal and air duct detectors, manual stations and up to five photoelectric or infrared detectors on each zone. The CTZ system can sound the alarm, close smoke doors, activate extinguishing systems, shut down equipment, etc. The CTZ Panel may be used with Remote, Power Failure, and Leased Lined Annunciators.
RELEASE SMOKE DOORS
ACTUATE EXTINGUISHMENT
START OR STOP EQUIPMENT
*Max. of 5 infrared and photoelectric detectors
Self-Contained Detector.
The Pyr-A-Larm type FDU fire detector unit can be used for upgrading existing systems with an ionization, thermal photoelectric, or flame detector capability when only one detector is required.
Also the FDU fire detector unit can be used for releasing system (e.g. in vaults), controlling equipment, etc., when only one ionization, thermal, photoelectric, or flame detector is required.
MODEL FDU
IONIZATION OR FLAME DETECTOR
EXISTING FIRE ALARM SYSTEM
ACTUATE EXTINGUISHMENT

IX.

Fire in the Library, 1973-1978

"As the contents can never be made incombustible, there will always be the risk of fire in libraries quite apart from the construction of the building . . ."

Cornelius Walford, at the Second Annual Meeting of the Library Association of the United Kingdom, September 1879

The San Diego Aerospace Museum and Library, February 23, 1978

FIRE IN THE LIBRARY, 1973–1978

The library of St. Clair College of Applied Arts and Technology at Windsor, Ontario burned during the night of November 12, 1973.[1] A faulty fire alarm system figured in this $303,000 loss. Although the alarm signal was heard in the building, the circuit to the annunciator panel had been disconnected. This meant that the fire department arriving on the campus could not determine the location of the fire when they checked the annunciator. There was a considerable delay, and the delay increased the destruction from the intense heat and heavy, black smoke.

Laminated counter tops peeled; light fixtures and phonograph records melted. Head Librarian Anita Blair described the grim picture painted by fire in her library: "I will never forget the sight when I walked into our once colourful library. The total absence of colour was shocking. A high intensity light was shining on jagged remnants of shelves at the source of the fire. Firemen were floating around in long black coats and even their faces were black. Everything that could be seen was black."[2]

Careful investigation ruled out electrical failure and arson. Investigators decided that a cigarette may have been negligently left in loose papers stored in cardboard boxes, possibly the only materials in the library that could have caught fire in this way.[3]

Many books were salvaged at a bindery by removing the covers and leaving them exposed overnight to an air-chemical mist that neutralized the smoke odor. The library was out of service for six months, and another year went by before all the books and catalog cards could be replaced.

On August 7, 1974 smoke spread through the Alderman Library at the University of Virginia early in the afternoon. The first fire department unit arrived within three minutes, but a total of 33 volunteer and paid firemen worked the fire and were on the scene more than 5 hours.[4] Fire had started in newspapers in the basement level, creating dense smoke which was circulated by the air handling system throughout the library. The absence of windows made venting of smoke difficult, and it was some time before the seat of the fire could be located.

Also in August of 1974 Newport High School in Bellevue, Washington lost its learning resources center in a night fire set by four juveniles, the oldest 18. Beer bottles filled with gasoline were ignited and thrown into the building. The group had come from a party where drugs and intoxicants were available. The loss was more than $1,000,000, including 40,000 books, 10,000 periodicals, $50,000 in audio-visual equipment, as well as the destruction of an entire wing of the building. The ringleader of the arsonists, a maladjusted former student, was sentenced to five years' detention and work-release rehabilitation; the younger boys were placed on two years' probation.[5]

At Spring Valley, New York, the Martin Luther King, Jr. reading room burned during the night of October 13, 1974.[5] Housed in a portable building, the library had more than ten thousand items of black literature, many of them concerned with the life and achievements of Dr. King. The collection was almost totally destroyed. Arson was ruled out after a careful investigation. The volunteer fire department had arrived promptly, summoned by an automatic fire detection system, but there was little water at the nearby hydrant, not enough for an effective hose stream. The library has since been relocated in Spring Valley's Martin Luther King Community Center.

The Kent Branch of the Toledo-Lucas County Library System was heavily damaged by fire the night of December 1, 1974 at Toledo, Ohio. The fire occurred during a blinding snowstorm; it was thought to be the result of an electrical fault. Insurance claims totaling $354,848 were paid. The damaged building was abandoned and a new structure was built for the library across the street.

Incendiary Fires

An incendiary fire struck the main building of the New Rochelle, N.Y. public library during the night of January 27, 1975.[6] Damaged or destroyed were between 8,000 and 10,000 books, most of which had to be replaced. The loss was set at $85,000, which covered cleaning and repairs to the building and replacement of books. Portions of the book stacks were burned out. Water damage from firefighting operations was incurred in bookmobile collections in the basement.

Windows had been smashed at the rear of the building and "flammable substances" thrown in and ignited. A neighbor reported the fire at 12:35 A.M., and the fire department was busy until 3 A.M. putting it down. Emergency repairs and cleaning were started by the staff at once with the help of various city agencies and commercial establishments. Only two days later the library was reopened on a limited basis.

Ten days after the New Rochelle incident, on the night of February 6, 1975, the Tufts Library at Weymouth, Massachusetts was set afire by vandals.[7] Breaking in through doors and windows at the rear of the library, they set fires in two bookshelves in the children's section. An automatic (rate-of-rise) heat detector system sent an alarm just after midnight, and the Weymouth fire department responded. Working with three hose lines, they were able to confine the fire to the room where it started. Damage was held to $59,000, due to good salvage work and judicious use of water by the firemen.

Students and janitors joined forces with Chicago firemen March 17, 1975 to put out a series of small fires on four levels of the Regenstein Library at the University of Chicago.[8] Stacks of books and pamphlets were used for starting the fires. Police went through the stacks with tracking dogs, and although they failed to locate the arsonist, there were no more fires.

Fireworks and Arson

Arson with fireworks took place May 19, 1975 at Scarborough, Ontario. Someone climbed a low roof at the Bellmere Junior Public School, broke an upper window of the library and fired a Roman candle into the draperies. These caught fire and in turn ignited the raw cedar interior finish, reference bookshelves and carpeting. A large globe mounted on casters burned explosively and contributed heavily to the troublesome black soot and plastic residues that covered everything. The loss was estimated at a nominal $12,000, but the library was not restored to service until September.[9]

Smith College at Northampton, Massachusetts suffered a $275,000 loss when the reference room of the Neilson Library burned during the night of October 21, 1975.[10] The entire building had been equipped with a products-of-combustion detection system except for the reference room, which had somehow been omitted from the plan. Smoke from the fire was carried through the air handling system into another room, where a sensor relayed an alarm to the fire department at 3:55 A.M. Firemen held the fire to the reference room. The furnishings and interior finish were largely destroyed and there was extensive damage to books.

The attic and newspaper rooms in the basement were protected by automatic sprinkler systems. Plans were made to extend the automatic detection system to include the entire building, and to provide similar protection for a new library addition then being planned.

On a Saturday morning in the fall of 1975 teen-age boys pushed burning paper through the mail slot of the New England Deposit Library in Boston, adjacent to the Harvard Stadium. Firemen responding to an alarm were able to limit the damage to the burned-out entry. Robert Walsh, a Harvard library administrator, waited hours for workmen to come to repair the door, then nailed a slab of plywood over it himself.[11]

January 12, 1976 three pipe bombs were discovered planted beneath the Dag Hammerskold Library at United Nations Headquarters in New York City.[12]

Burned Up: One Million Books

One of the largest book losses to fire occurred at Austin, Texas February 2–3, 1976.[13] The fire destroyed a bindery and warehouse for school textbooks. Bindery machinery and equipment and one million textbooks went up in flames. The Texas Education Agency lost $5 millions in books, while the building loss was one million. There was no automatic fire protection system of any kind.

A telephone call alerted the Austin Fire Department at 11:35 P.M. They responded even though the fire was outside their response area. Their officers then coordinated the efforts of volunteer departments responding along with the Austin firefighters. They held the fire in check for a time, but the area where the books were stored could not be saved. Adjacent enterprises endangered by the fire were a lumber yard and three large warehouses. Volunteer fire companies, including a number of lady firefighters, kept a long vigil with hose streams to protect these properties and prevent the fire from getting into the high desert grass in the area.[14]

The Education Agency was not insured; the huge loss was assumed by the state legislature through an appropriation. The building was rebuilt with somewhat better fire divisions. The Agency, a tenant in the building, was considering the economics of automatic protection for the books.[15]

The Walker, Minnesota, Public Library had the distinction of being the smallest Carnegie library in America. It was one unit in the Kitchi Gami Regional Library System, which includes two Indian reservations.[16] On the night of March 31, 1976 it burned to the ground. Just before 7 P.M. a dull explosion in the basement signaled the presence of the fire and an alarm was sounded. Eight hours later the fire was under control, but only remnants of the brick walls remained; the library was gone. A large fuel tank in the basement burst about an hour after the first alarm, doubling the intensity of the blaze and ending any hope of saving the books.

The loss was estimated at $160,000, half for the building and half for the furnishings, equipment and collections. There was no suspicion of arson. The insurance carrier sued to recover from one or another parties concerned with supplying fuel oil and installing the fuel oil filling piping, and the litigation was in progress two years after the fire. Plans were underway for building a new Walker Library either at the same site or at another site more suitable for serving its patrons. A temporary library was set up in the basement of the City Hall.

Incendiary fire destroyed the Burroughs High School library at Ridgecrest, Kern County, California on September 4, 1976, three days before the opening of the new school year. The fire was reported at 4:03 A.M., and was under control two hours later. Flames rose eighty feet from the one-story brick and concrete block building. Eight of the nine rooms in the library were burned out. The loss was set at one million dollars, $175,000 of which was the cost of the books destroyed. Investigators said the fire was definitely arson.[17]

Kern County had experienced a library burning before this. On November 20, 1974, while patrons were in the library early in the evening, someone dropped a lighted book of paper matches into the book drop at the Tehachapi Branch.[18] Burned were the foam rubber liner, the box itself, the front door of the library and one side of the card catalog. In spite of the fire the library remained open to patrons until the usual closing time.

Langley Hall at the University of Pittsburgh was shaken by a violent explosion January 20, 1977.[19-20] Natural gas from an underground main or supply line had apparently seeped through pipe trenches or porous soil into the basement of the building. Two persons were killed. An electron microscope that weighed one ton was moved fourteen feet by the explosion. The life science and psychology library was extensively damaged. Books wet from firefighting quickly froze; 3000 of these were placed in a freezer to await restoration. Conservators using modern cleaning and drying techniques were able to save most of the books.

Engineering Library, University of Toronto

Only three weeks later, on February 11, 1977, came the destruction by fire of the Sir Sandford Fleming Building at the University of Toronto, which contained the Engineering Library.[21] The fire burned undetected for some time in concealed spaces of the old building, then was reported at 2:21 A.M. by students who smelled smoke. An earlier alarm may have been sounded by an automatic system protecting computers in the Burton Wing of the building, but that system was being tested in a maintenance routine at the time, and this confused the situation. There was no alarm system protecting the part of the building occupied by the Engineering Library.

The battle of the Toronto Fire Department to save the building took on epic dimension as the hours went by. The first arriving fire company found a well established fire in the auditorium, adjacent to the library, and the officer in command sent in additional alarms. Ninety-one firemen fought to stop the fast moving fire, which traveled behind a false wall, over a suspended ceiling, through an unused duct into the attic, then down again into various parts of the building.

More than a mile and a half of fire hose was used to pour water into the building over a period of seven hours. Firemen placed 248 salvage covers over bookshelves to prevent water damage to books—six acres of canvas—probably the most prodigious conservation effort of this kind ever accomplished. Eight firemen were injured, several when the roof fell in and caused a "blowout" of flying debris.[22]

Weather conditions were in sharp contrast to the 91°F. and 41% humidity that prevailed when the Temple Law Library burned in Philadelphia July 25, 1972 in daylight. The Toronto fire was fought in a chilly early morning at ten below zero. This bitter cold produced one benefit, in that books that were soaked were at once quick-frozen, preserving them from further deterioration for the time being. Of these wet and frozen books, 500 that were judged rare or hard to replace were put through a modern salvage routine in vacuum chambers at the Canadian Forces Base, Downsview, in accordance with the University's Disaster Contingency Plan.

Totally destroyed in the fire were University papers, photographs and the personal library of L. E. Jones, Professor Emeritus, curator and historian of the Faculty of Engineering. A number of items dated back to the 1700's. Out of 80,000 other items exposed to damage by fire and water about 4000 monographs, 3,000 case bound books and 5,000 papers or unbound journals were destroyed. Of the remaining 68,000 about 5,000 were considered damaged but repairable; another 5,000 were either beyond repair or of insufficient value to justify the cost of repair. The loss directly related to replacement and repair of books was set at $600,000.

Disaster Plan Put to Work

Chief Fire Prevention Officer Herbert F. Gladney credits the Disaster Contingency Plan for University of Toronto libraries with mitigating greatly the destruction of the collections by insuring efficient operations for salvage of books.[23] Using experience gained in the Philadelphia fire-and-water disaster of 1972 and a similar incident at Overland, Missouri in 1973, the University's plan was prepared in 1976. William H. Adamson, adjustor for insurance companies concerned, worked with University authorities to perfect the details.[24] Listed on charts were 20 commercial establishments pledged to immediate loan of services or facilities in an emergency, as well as specific assignments of duties to physical plant and library officers and staff.

Final determination of the extent of losses from this fire awaited the reconstruction of Sir Sandford Fleming building; preliminary estimates placed the loss at between $4 million and $5 million dollars. The library was to be moved to the Toronto Central library until new construction was completed, the city library having been relocated in the new Toronto Metropolitan.

The disaster increased the University's awareness of the threat of fire and gave substance to arguments for improved fire protection, particularly for automatic systems for fire detection and suppression in historic buildings. The new University complex of buildings for athletics was scheduled to be provided with automatic sprinkler systems, an indication of a stronger policy on protection from fire.

Late at night on March 15, 1977 the Rhodes Memorial Library at Gideon, Missouri was burned out by a fire that started in an adjacent laundry.[25] Children wept watching the fire; 125 of them in a town with a population of 1112 were enrolled in a summer reading program. When the fire was out about half the books were soaked or otherwise damaged. Two days later the library was back in operation in another location down the street, checking out damp and smoky books. Mayor Johnson declared that they would waste no time in rebuilding, and would not depend on state or federal funds to do so. Librarian Cora Hutchison reported that the busy library circulated 14,000 volumes in 1976, indicating a very high level of use by the community. Insurance on books was judged insufficient and was to be increased at the first opportunity.

Teen-Age Arson

The Delta Pioneer Library at Delta, British Columbia was destroyed March 21, 1977, when it was fire-bombed by three teen-age boys.[26] They placed an incendiary device in the book drop early in the afternoon, and within seconds the building was ablaze. The fire department had immediate notice of the fire and responded at once, but were met with a roaring fire. By the time they got it under control, the whole main section of the building was gone, and with it 12,000 books, 14,000 paperbacks and 3,000 periodicals. Damage to building and equipment was estimated at $200,000, and to the collections, $75,000; about half the loss was expected to be recovered from insurance.[27]

INCENDIARY FIRE IN LIBRARIES—A CHANGING PICTURE

(A) Political, Racial or Religious motive:	*(B) Spite, personal vengeance or other "emotional" motivation:*	*(C) Casual vandalism by males 12–20:*
1914 suffragettes in Northfield, England		
	1950 Government Dept. library at University of Michigan	
	1951 State Library, Michigan	
	1959 Bayonne, N.J., P.L.	
1966 Jewish Seminary, N.Y.C.		
1968 Holyoke, Mass. (Community college)		
1968 Middle River, Maryland		
1968 US Info. Svc. Brazil		
1968 US Info. Svc. Brussels		
1969 US Info. Svc. Frankfurt	1972 Long Beach, Calif.	
1969 Amer. Memorial, W. Berlin	1974 Centerville, Fremont, Calif.	
1969 Indiana University (twice)		1972 San Francisco, Ortega Branch
1969 Bronx NY Univ. (Gould Lib.)		*1974 Seattle, Fremont Branch
1969 Ropkins Br., Hartford, Ct.		*1974 Kern County, Calif., Tehachapi Branch
1970 Univ. Washington (Suzallo Lib.)		1974 Bellevue, Wash., Newport High School
1970 Univ. California (Doe Library)		1975 King County, Wash., Federal Way Branch
1971 Univ. Washington (Engr. Library)		1975 King County, Wash., Carnation Branch
1971 Univ. Wisconsin		1975 Boston, Mass., N.E. Deposit Library
1971 L.I. College of Medicine		1975 Scarborough, Ont., Bellmere School Library
1971 Tufts Univ., Fletcher Lib.		1975 University of Chicago, Regenstein Library
1972 US Info. Svc. Dacca		*1975 New Rochelle, NY, Public Library
		*1975 Weymouth, Mass., Tufts Library
		*1976 Ridgecrest, Calif., Burroughs School Library
		1977 Delta, B.C., Canada, Pioneer Library
		1978 San Diego Aviation Museum and Library
1976 United Nations, N.Y.C. attempt at bombing		*1978 Crane Br., Buffalo-Erie County, N.Y.
		*1978 South Avenue Br., Rochester, N.Y.
		1978 Woodhaven Br., New York City
	1978 Youngstown, Ohio (motive uncertain)	*1978 Riddle Br., Douglas County, Oregon
1978 Memphis, Tenn. strike by firemen		*1978 Concord, Calif., High School Library

* This incident is *believed* to have been perpetrated by juveniles. The other entries in this column were based on reports of witnesses, arrests of the arsonists, or other substantial basis.

Along with "rows and rows of blackened books which disintegrate at the touch" some irreplaceable items were lost. Historical records of the Delta area were destroyed, as well as audio visual equipment, films and files relating to membership and circulation. The prized Winston Churchill collection, bound in cream colored leather and at first thought to be intact, was found later to be discolored and with bindings damaged. A recommendation was made that local libraries be equipped with a "fireproof book deposit" to prevent another such disaster, and insurance coverage on books was increased from $1.50 to $3.00 per volume.[28]

A fire in the Grosse Pointe, Michigan, central library on April 9, 1977 actually extinguished itself.[29] Starting during the night in an electrical panel in a janitor's closet, the fire caused considerable smoke damage throughout the library. Before it was discovered, however, the fire was extinguished by a spray of water from a 1½ inch copper pipe joint that opened in the heat. Police officers found water running out the door at 5:30 A.M.

Available Water Sprinkler Protection

Damage to the building and the books from water and smoke amounted to $47,000. During repairs a fire door was installed to improve the compartmentation of the building. This incident suggests the very practical measure of spot sprinkler protection for janitor closets and other locations where the fire hazard may be recognized as higher than other areas in the building. A single sprinkler head can be installed and connected to any reliable water supply. The Life Safety Code, NFPA 101(1976: 9-3.5.2) *requires* this arrangement in schools, with the proviso that water for as many as six sprinkler heads may be supplied from the (existing) domestic water supply system as long as the supply is adequate to provide water at the rate of .15 gallons per minute for each square foot of floor area.

At Ceres, California, in Stanislaus County, the Gondring Library was burned out in an arson incident August 14, 1977. Two boys, ages 14 and 16, who were "just fooling around," dropped a single paper match into the book drop. The resulting fire caused more than $200,000 damage to the library and the city hall complex, built only three years earlier.

The fire damaged a work room, offices and other rooms. Fire department officers, skeptical about the story that a single match started the fire, carried out trials with a simulated book drop with appropriate contents. They learned that more than half their attempts started fire with a single match!

Fire Marshal John F. Weber of Ceres recommended (1) elimination of book drops opening inside buildings; (2) a central station early warning system to summon the fire department; (3) an automatic suppression system; to which he added, with the cynicism of a veteran public servant, "The final alternative . . . is to do absolutely nothing, hoping that such a fire never occurs again."[30]

Book Drops Can Be Protected

There are three book drop plans for improving the fire risk. Metal book carts are available in various sizes, designed to be fire resistive and to receive materials at the book drop. Another method is the free standing book drop box placed outside the library. One such box is designed for curb service and is called a "drive-up book collector." At Ceres and elsewhere in the Stanislaus County system, libraries have closed off their mail slots and book drops and are using the outside boxes. Another method is to reinforce the existing book drop with fire resistive materials, as one Minnesota library has done with concrete block construction, creating a small vault to contain any fire.

A gas-fired space heater was the apparent starting point of a fire that burned out the Regional Library Annex in Statesboro, Georgia on October 7, 1977.[31] The loss of books and equipment was estimated at $70,000. Firemen had the advantage of familiarity with the premises, gained in an orientation visit only a few days before the fire occurred.

The loss included 9,671 books completely burned and 645 cassettes, films or film strips damaged or destroyed. Film strip boxes preserved their contents well, but cassettes failed to survive the heat. Microfilm in plastic cases and metal cabinets survived, but the film required cleaning. Local history materials in metal file cabinets survived, and only envelopes and file folders needed to be replaced.[32] Only a few days after the fire a bond issue was voted in, assuring the funding of a new facility.[33]

The San Diego Aerospace Museum and Library burned in a catastrophic fire apparently set by juveniles the night of February 22, 1978.[34] Estimates of loss ran as high as $15,000,000 in museum artifacts, works of art, books and other library materials and equipment destroyed, as well as loss of the structure. A special collection of more than 10,000 volumes, the Prudden Historical Aviation Library and Archives, was totally destroyed.

The Aerospace Museum and Library were located at Balboa Park in the Electric Building, which got its name from its original function as part of the 1915–16 Panama-California Exposition. This was a very large building of open construction with a false ceiling 29 ft. above the floor. The basic construction was of timber, wood lath and plaster, and some of the plaster had been stripped away in a reconstruction project; slabs of plywood covered the wood lath, and in the plywood arsonists set the fire.[35]

No Duplicate Card File

Fire companies responding to an alarm had little opportunity to do any effective fire fighting. The flames soared hundreds of feet into the night sky. Lost were ninety percent of all the books, periodicals, films, photographs, and original works of art related to aviation and space science. Fifty thousand catalog cards were lost, and there was no duplicate card file. Four large boxes of wet library materials were placed in cold storage to wait for restoration in the vacuum chamber process, from which they were later returned in good condition. Library archivist Bruce Reynolds recommended stronger construction and automatic sprinkler protection for the library.

Only a few days after this disaster another set fire in Balboa Park burned out the Old Globe theatre during the night, a few hundred feet distant from the first fire. While the auditorium was gutted, the stage block remained relatively undamaged by fire, although drenched with smoke and soot. The automatic sprinkler system required by law on the stage prevented the flames from reaching the stage settings and draperies. A few sprinkler heads on the proscenium wall opened to control the fire at that point.[36]

The Crane Branch of the Buffalo-Erie County Library System was badly damaged in a fire set by vandals the evening of April 27, 1978. They entered a rear window and pulled open catalog files before setting the fire, which caused approximately $27,000 damage to books and equipment and $50,000 to the building.

The South Avenue Branch at Rochester, New York was set afire, apparently by a burglar, in the early morning hours May 23, 1978. The damage was estimated at $140,000, and the building was damaged beyond repair. A few books on local history were sent out for restoration.

The Riddle, Oregon branch in the Douglas County system was damaged by fire started in a book return in the afternoon of June 6, 1978. The fire smoldered for some time before being discovered. Volunteer firemen spread salvage covers over books before throwing water into the building, but the collections were thoroughly steamed in spite of this. A shipment of books went to Jeff Brown at Lowie Museum on the campus of the University of California at Berkeley for fumigation and repair. Damage to the building amounted to about $30,000. Although children were observed playing with matches on an adjacent playground the day of the fire, no one saw the person who dropped a match or burning paper into the book drop. Douglas County has abandoned inside book drops.

The Woodhaven Branch of the Queensborough Library in New York City was burned out by boys who broke in to steal petty cash June 17, 1978. The fire was set in an upstairs office where they had found $3 or less. Three boys, the oldest age 15, battered a back door and broke out a wood panel to get in. This was the third such break-in this year. Smoke damage to the books was severe; the total damage was estimated at $250,000.

The Vance Avenue Branch library in Memphis, Tennessee was destroyed by fire July 1, 1978 during the strike by Memphis firemen. The loss included 22,000 books, furnishings and equipment valued at $15,000, and the total loss of the building. It is expected that the library system will rebuild at another location. The Vance Avenue Branch was the first in the system occupied as a separate branch, being established in 1938.

Police called to investigate a break-in saw fire in the vacant building next to the library and only about 25 ft. away. Auxiliary firemen responded after 30 minutes and attempted to control the fire with an aerial tower, while striking firemen stood by and watched. The tar-and-gravel roof of the library portico smoldered in the radiated heat from the burning building, then burst into flame explosively, and the library was doomed. The library was of wood frame construction with stone veneer and cement stucco exterior. This was only one of several hundred incendiary fires set during the strike, many of which were attributed to the striking firemen.[87]

On the morning of October 2, 1978 the main library of the Youngstown-Mahoning County Library at Youngstown, Ohio was struck by an incendiary fire.[41] A custodian coming to work before 6 A.M. saw flames in the top (6th) floor and hailed a passing patrol car to report the fire. The fire department response was prompt; fire damage was held to the 6th floor, but there was considerable smoke and water damage to books and premises in spite of the use of salvage covers.

Investigators found that 4 separate fires had been started. Although extensive repairs would be needed, the library was open to patrons again October 10. Losses included government documents, historical scrapbooks, miscellaneous files and furniture used to stock branch libraries.

The library of the Concord, California, High School was severely damaged by a night fire November 10, 1978. The fire was discovered and reported by a night custodian; four engine companies and 22 firemen responded, and the fire was subdued within 30 minutes. Damage to books, library materials, interior finish and equipment was severe and smoke damage was heavy. Fire had been set at several places in the library. The loss was estimated at $120,000.

Arson: An Important Concern for Libraries

Arson is considered to be the nation's fastest growing crime.[88] The National Fire Protection Association reported that in the decade ending in 1974 the number of incendiary and "suspicious" fires in the U.S.A. rose from 30,000 to 114,000, a 237% increase, then another 30,000 in 1975, or 25% in one year. While the number of libraries struck by incendiarism has been a small portion of the total, it has been an important share of the total number of fires in libraries.

There is a pattern in arson committed against schools, churches and libraries which is quite different from commercial and industrial burning for profit. As shown in the chart (page 99) libraries are generally believed to be set afire with one of three types of motives, and these apply also to schools and churches. They are (A)racial, political or religious motive; (B)spite, personal vengeance or other "emotional" motive; or (C)casual vandalism by teen-aged boys, apparently without any conscious motive.

Some of the observations concerning industrial and commercial arson seem to apply to libraries as well. A report by an insurance officer[39] described a study of 736 arson fires over a 6-year span. Properties found especially vulnerable were "Facilities which serve the public and include combustible storage," for example. Also, "The source of ignition was not so important as the availability of fuel in an unattended or unsupervised area . . ." The study found that arson was committed by "intruders, employees, visitors and juveniles," and in that order of frequency. Our tabulation does not conform to this pattern, in that juveniles have lately come into prominence in this sorry crime of library arson.

There are some fairly simple and effective ways of reducing the likelihood of incendiarism. Here is a list, commencing with the least costly:

(1)Reduce opportunities for casual visitors to go to unsupervised areas in the library;

(2)Keep the library lighted at night;

(3)Seal book drops and mail slots, or provide containers or carts designed to receive materials through these openings and contain a fire;

(4)Provide perimeter lighting for dark areas at the rear of the building and other areas not readily observed from the street;

(5)Strengthen doors and windows at the rear of the building and in other areas not readily observed from the street;

(6)Install an intrusion alarm system;

(7)Install a fire (and smoke) detection system communicating directly to a fire department or a central communications station;

(8)Install a fire suppression system with central station reporting capability.

Building design directly contributes to security against vandalism when the nature of the crime is considered in the planning. The Denver Public Library obtained several advantages with a nearly windowless plan, not the least of which was the discouragement of vandalism.[40] The least costly of the recommended changes in the list above, closing off book drops and mail slots, would have prevented five of the thirteen incidents listed. It is clearly futile to consider doing very much to anticipate or influence the perverse behavior of maladjusted adolescents or other persons except to strengthen the physical defenses of the library against attack.

REFERENCES

1. Windsor Fire Department Report of November 13, 1973.
2. Letter from Anita Blair, April 14, 1976.
3. Conversation with James Scott, Principal, Newport High School.
4. Charlottesville Fire Department Incident Report, August 4, 1974.
5. WILSON LIBRARY BULLETIN, January, 1975:399.
6. New Rochelle Public Library news release, February 4, 1975.
7. NFPA Fire Analysis Department memorandum, July–August, 1975.
8. LIBRARY JOURNAL, vol. 100, no. 12, June 15, 1975:1172.

9. Conversation with Ken Turton, Principal, Bellmere Junior Public School.
10. Letter from Kenneth Smith, Safety Officer, April 5, 1976.
11. AMERICAN LIBRARIES, vol. 7, no. 2, February, 1976.
12. LIBRARY JOURNAL, vol. 101, no. 2, June 15, 1976:299.
13. FIRE ENGINEERING, vol. 129, no. 8, August, 1976:71.
14. Austin Fire Department Reports of February 3, 1976.
15. Letter from J. Henry Perry, November 9, 1977.
16. Conversation with Marlys O'Brien, Director, Kitchi Gami Regional Library System.
17. BAKERSFIELD CALIFORNIAN, September 5, 1976.
18. Letter from Nina Caspari, Kern County Library System, October 31, 1977.
19. PITTSBURGH POST GAZETTE, February 12, 1977.
20. PITTSBURGH PRESS, February 11, 1977.
21. TORONTO STAR, February 11, 1977, also February 12 and 14, 1977.
22. Toronto Fire Department Report of February 11, 1977.
23. Letter from H. F. Gladney, October 6, 1977.
24. Letter from W. H. Adamson, May 19, 1977.
25. Sikeston, Missouri DAILY STANDARD, March 20, 1977.
26. Delta, British Columbia OPTIMIST, March 21, 23, 30, 1977.
27. Letter from Cecelia Duncan, Community Librarian, Delta, February 2, 1978.
28. Letter from W. H. Overend, Director, Fraser Valley Regional Library, February 8, 1978.
29. Grosse Pointe, Michigan NEWS, April 14, 1977.
30. NEWS NOTES OF CALIFORNIA LIBRARIES, vol. 72, no. 1, 1977:36–40.
31. Statesboro, Georgia HERALD, October 7, 1977.
32. LIBRARY JOURNAL, vol. 102, no. 22, December 15, 1977:2463.
33. Letter from Isabel Sorrier, Director, Statesboro Regional Library, January 12, 1978.
34. San Francisco CHRONICLE, February 24, 1978.
35. Letter from Bruce Reynolds, April 6, 1978.
36. Conversation with Ed Carey, Fire Inspector, San Diego Fire Prevention Bureau.
37. Conversation with Lamar Wallis, Director, Memphis-Shelby County Public Library and Information Center.
38. "Fastest Growing Crime—Arson," United Press, June 5, 1978.
39. BUSINESS INSURANCE, June 30, 1975:16.
40. ARCHITECTURAL RECORD, vol. 156, no. 5, October, 1974:38.
41. Conversation with Jeanne Dykins, Public Relations Officer.

1.

2.

3.

4.

University of California Libraries

1. Santa Barbara
2. Santa Cruz
3. Irvine
4. Riverside

Library Risk Management: Current Topics

"The speed with which fires grow is seldom understood by the general public and delay in calling the brigade is all too frequent. There is a real need for more in-built protection to keep fires within reasonable bounds until the fire-fighters are able to take over. Much more could be done at the various stages of designing and equipping a building to confine fire in this way . . ."

Fire and the Architect, 1976, Fire Protection Association and the Royal Institute of British Architects

The Metropolitan Toronto Library, Canada

The Metropolitan Toronto Library is centered around a spectacular atrium with fountains and glass-enclosed elevators. It is completely protected by fire suppression and detection systems and by other fire defenses to provide a high level of life safety as well as protection for the collections.

LIBRARY RISK MANAGEMENT: CURRENT TOPICS

The Systems Approach

A modern concept important in the fire safety of new libraries (and other buildings) is the *systems approach* to fire protection in building design. This concept derives methods from system safety and fault tree analysis as used in aviation and aerospace industries during the 1960's to achieve a high degree of safety in complex systems. In the early 1970's Irwin A. Benjamin and Harold E. Nelson were leaders in the adaptation of these methods to fire protection planning for structures. Since 1971 the National Fire Protection Association has had a *Committee on Systems Concepts for Fire Protection in Structures.*

In the systems approach fire safety is given equal consideration in the design along with other subsystems such as the structural, mechanical or electrical subsystem. The defense against fire is worked out not simply in compliance with specification or performance codes, but according to engineering technology and methodology. The results are proving satisfactory not only in fire protection but in obtaining cost-effective construction.[1]

Systems fire safety planning is a goal-oriented process in which a fire problem is worked out by the fire protection engineering consultant and the client library. The eventuality of fire is a working hypothesis. The client librarian, director or governing board representative must specify *the extent of damage from fire the library is prepared to accept.* The consultant can then respond by designing a control system to meet this specification through such control measures as confinement, detection and suppression. The systems approach reduces fire in the library to a designable physical and chemical problem.

Closely associated with the systems approach to fire safety in structures is the device known as *the decision tree* (page 122). This is a diagram in which the objectives to be attained in safety from fire are shown at the very top of the tree, while the lines and boxes radiating from the top define the methods for attaining the objectives. Numerous alternatives can be shown in such a diagram.

The decision tree becomes a document for constructive thinking and planning, as well as an instrument of discipline for the architect and the engineer. It makes it relatively easy for the fire protection consultant to discuss fire protection alternatives with the architect, and for the architect, in turn, with the librarian or the governing board. Actually, the decision tree is only a pictorial representation of the reasoning process followed by fire protection professionals for some time.

Standard Fire Defenses of a Library

Fire protection consultant Rex Wilson[2] sees a standard doctrine developing for the fire protection defenses of a well-equipped library approximately as follows:

(a) A reliable products-of-combustion automatic detection system, reporting to a central station, preferably a fire department, in addition to any local alarm sounded in the premises protected;

(b) A Halon 1301 system or other approved prompt action system using an agent other than water whenever the greater cost can be justified by the extra-high values or special character of the collections;

(c) A system of automatic sprinkler (water) protection as a secondary line of defense. Operation of this system is contingent upon failure of the alarm (of the detection system) to bring to bear effective first-aid (fire extinguisher) suppression. When used in concert with (b), above, it provides in-depth protection also in the event that unusual circumstances prevent the total success of that system.

(d) These defenses (a, b and c) apply in addition to the intrinsic fire protection design features of compartmentation, control of fuel loading contributed by furnishings, equipment and interior finish, safe air handling design, and other elements.

Automatic sprinkler protection is now seen as *the firestopping system.* It takes command only after an appropriate delay to permit local attack on a fire, and then operates to limit the fire to the fire zone where it originated. If the detection system's early warning fails for any reason to bring about effective and timely action by the library staff or the fire department, the sprinkler system acts to discharge water directly over the fire with a single head, and with other heads opening only if the fire is sufficiently hot and fast-burning.

The New Libraries

In the discussion of new libraries in this section and improvements in existing libraries in the next it has been possible to include only a very few. Many other noteworthy new libraries and many meritorious programs of improvements might have been included.

Toronto Metropolitan Library

Modern design in library buildings is atrium style, with a lofty hollow space in the center around which the activity of the building takes place. In any very large building the atrium has somewhat the breathtaking spaciousness of a cathedral. This is doubtless what Barry V. Downs had in mind in commenting on the new library at Toronto[3] when he wrote "Next to houses of worship, libraries are the most ethereal spaces we inhabit these days."

The Toronto Metropolitan Library, first opened November 2, 1977, brings together as many as 1305 users and over a million books in the most pleasant of surroundings. It was designed with such people-pleasing amenities as quiet-running glass walled elevators, three pools and a waterfall. But it also has an outstanding pattern of systems for fire protection and life safety.

There is a complete automatic sprinkler system, fitted with conventional heads in public and administration areas, but with stop-and-go heads in book stacks. Rare books and special collections have automatic suppression systems using Halon 1301, the gas extinguishant. Interior finish materials have a flame spread rating of 25 or less, virtually noncombustible.

One detail serves to illustrate the careful attention paid to control of the burning characteristics of materials. Designer Stephen Hogbin's "extended hours screen" is a divider for separating the main reading area into two parts, one of which may be used after regular library hours. It was built of 350 red oak planks, each 2" by 2" by 7 ft long, leaning either way from the base to simulate a wind-battered fence. Each piece was given four coats of finish, two of them fire-retardant, making the barrier quite resistant to fire and well within the stringent flame spread limitations required for acceptability.

Other features for fire safety include an annunciator system to pinpoint the location of any fire, an exhaust system to extract smoke from the building, pressurized stairwells to inhibit smoke entering and keep them open for the use of patrons, and a limited detection system for early warning of fire. An emergency generator stands ready to assume the electrical load if the normal power supply should be interrupted. Finally, generous exit facilities were provided on the basis of modern exiting time analysis research, permitting simultaneous evacuation of all levels in the emergency.

An important design consideration was the accommodation of the physically handicapped. *Canadian Architect*[4] credits Raymond Moriyama with doing this well in all his buildings. In this one he provided extension legs on tables to allow for wheelchairs, kept desks and inquiry counters low, and in various other ways applied design skills to make the library a congenial place for persons with impaired mobility.

Not far away in Hamilton, Ontario, is the new Hamilton Central Library,[5] still under construction. It, too, will have a handsome atrium, as well as fire protection systems not unlike those at Toronto. There will be a complete automatic sprinkler system, smoke sensors throughout, and a capability for exhausting smoke during a fire. Special collections will be protected with Halon 1301.

Shower Wall Construction

An earlier atrium building is that of Bobst Library at New York University, in New York City, 200 ft square and ten stories high. In this library the towering atrium is protected on the various floor levels by partitions of wired glass backed with a line of sprinkler heads. This arrangement is known as a "shower wall," and is accepted by the New York fire authorities as a heat and smoke barrier in certain occupancies. It is not a complete sprinkler system, of course, and is of limited value except for the specific function it performs in cooling and preserving the glass partition.

Bobst Library has a complete detection system using both products-of-combustion sensors and rate-of-rise heat detectors. Basement and sub-basement are equipped with an automatic sprinkler system. Emergency lighting and fire extinguishers are provided on all levels, as well as standpipe and hose protection. An annunciator panel in the entrance lobby is duplicated in the basement control room. All nine elevators are engineered to return to ground level in an emergency through the use of a control key.

On the Stanford University campus a new addition is the Cecil H. Green library, storing about 1.25 million books and accommodating 1300 students. Like all the Stanford libraries, it is completely protected with automatic sprinklers and has standpipe and hose stations on all levels. There are advanced systems for smoke detection, public address communications, sound masking and security.

The Green library forms the eastern element of a quadrangle that includes the J. Henry Meyer Memorial Library and the Cubberley School of Education building. Nearby are the Herbert Hoover Archives in the Hoover Memorial Building, a facility opened to the public in December, 1977. Collections stored there, valued at $25,000,000, are protected by automatic sprinklers and other systems.

A great library on the University of Texas campus at Austin is the Lyndon Baynes Johnson Library, built by the University and maintained by the federal government. Completed in 1971, it is an imposing structure of marble, nine stories tall and largely without windows. There is an early warning detection system and a complete automatic sprinkler system.

Two smaller libraries at opposite ends of the country were designed by Mitchell-Giurgola.[6] The Tredyffrin Library in Stratford, Pennsylvania is equipped with an automatic early warning detection system, wired to the fire department. The Condon Hall law library at the University of Washington in Seattle has a smoke detection system in the library proper and an automatic sprinkler system in the ground floor "student street" and adjacent areas.

Pre-Action Sprinkler Protection

Another sparkling new library in atrium style is that of Western Illinois University at Macomb. Here the choice was a pre-action system, the empty pipe system that charges itself with water when fire is detected by a smoke detection system. In non-finished areas upright sprinkler heads are mounted on exposed pipe; in finished areas the heads are flush-mounted types that are less visible.

Reading and stack areas are adjacent on the various levels. They are separated from the 5-story atrium by a 2-hour fire wall and Class B(1½ hr.) rated doors. Break-glass alarm stations are found on all floors. There are four exit stairways in 2-hr fire resistive enclosures, in addition to the stairway in the atrium. Special attention was given to the choice of carpeting, since in some areas it is used on walls as well as on floors, and therefore had to meet a stringent flame spread requirement.

Smaller libraries generally provide automatic protection with central station detection systems, and depend entirely on city fire departments for fire suppression. An exception is the Fremont Main Library in the Alameda County, California, system, a modern facility of modest size equipped with automatic sprinkler protection. Other libraries in the Alameda system lately have expressed interest in having book drops protected with a single automatic sprinkler head, a job to which the stop-and-go head is well suited.

British Public Records Facility

A recent report[7] gives details of fire protection in a very large and vital document storage facility in England. The Public Record Office at Kew, a suburb of London, houses the official records of the British Government and the Central Law Courts for England and Wales. Some documents date back to the eleventh century. They already occupy eighty miles of shelf space, and another mile of materials is added every year. The fire risk is very similar in many ways to that of a large library.

Replacing outgrown facilities in the center of London, the building was designed and built by the Property Services Agency and opened in 1977. It was decided not to provide a water system for fire protection, and the areas were much too large for effective application of any gas system. Instead, the designers relied mainly on tight compartmentation horizontally and vertically. The 4-story building is of reinforced concrete. There are two basements, one fitted with ionization type smoke detectors and the other with high expansion foam generators protecting repositories for documents.

One-hour fire rated construction is used to separate fire zones, except that the public staircase from the main entrance hall to the level above is provided with two separate wired glass partitions, each rated at one hour, while the space between is maintained free of combustible furnishings.

A documents conveyor system that penetrates the compartmentation barriers at various points in floors and walls is provided at every opening with fire shutters designed to close on the action of a fusible link. Emergency lighting is provided by lights of extra power to light up long passageways and provide focal points for evacuation of the floors during an emergency. Four escape stairways are pressurized, and upper floors have a smoke extraction system. Automatic systems are also operable by manual control. Air conditioning ductwork is equipped with automatic fire dampers at all points where ducts pass through a floor or wall, ready to shut off the duct on the action of a fusible link.

The new building and its systems were under the scrutiny of the Fire Research Station, and variations from standard fire-rated constructions were tested for acceptability. The building is to be continuously occupied, and there are constant security patrols, factors which have been given some weight in evaluating the fire risk.

James Madison Memorial Building

Scheduled for opening some time during 1979, the James Madison Memorial Building of the Library of Congress at Washington will be endowed with a formidable complex of fire protection measures. The fire problem here is one of very large dimensions quite literally. Each of the three stories below the Independence Avenue grade occupies nearly 8 acres, and the 6 stories above grade slightly more than 4 acres each. At the 7th level there is a penthouse for mechanical equipment for ventilation and other services.

There will be an atrium with one or more fountains and a garden. An exhibit hall is located just inside the monumental entrance from Independence Avenue. The massive exterior walls are without conventional windows except for a window wall at the 6th level. Below that point the windows are only 18 inches wide.

The fire protection plan for the Madison Building directly reflects the systems approach in design, and a primary objective was the confinement of fire to the fire zone of origin. The plan for achieving this is an in-depth array of fire defenses. First is traditional compartmentation, the reduction of very large areas into somewhat smaller areas with construction of rated fire resistance. Second is an early warning detection system that senses smoke or invisible combustion products and sounds an unusual "slow whoop" signal, at the same time sending a call to the fire department. At this stage a fire extinguisher or other first-aid firefighting device can be brought to bear on the fire, and information may be disseminated over the public address system.

The third defense is automatic suppression by an appropriate system. Heat from a fire has to reach 165°F. at ceiling level to open one sprinkler head and place water on the fire. When the fire is wet down and the temperature drops, the head shuts off the flow of water(being a stop-and-go type). In an area protected by the Halon 1301 system, the extinguishant gas floods the compartment protected and knocks the fire out instantly.

Fire protection is planned for 3 distinct occupancies: (1)general public areas; (2)Spaces vital to the continuity of services, like the control room and the computer rooms; (3)spaces housing irreplaceable collections. Occupancies (2) and (3) are due to have Halon 1301 protection, and may eventually have stop-and-go sprinkler protection also. This would provide 3 ranks of armament for fire defense.

In addition to these systems, the HVAC(heating, ventilating and air conditioning) systems have an important auxiliary role, the removal of smoke during a fire emergency. The fire zone can be isolated from the ventilating system through the use of motorized dampers in the ducts. Air is then exhausted 100 percent from the fire zone and supplied 100 percent to adjacent zones. This creates a pressure differential to keep stairways and other uninvolved areas free from smoke. Special ducts carry smoke away to the roof.

University of Riyad

Two great new libraries are planned for the environs of Riyadh, capital city of Saudi Arabia.[8] The University of Riyad will have a very large library in a massive complex of traditional Arab-style architecture nearly a mile long. The library will stand in the position of prominence at the point where two lines of buildings meet to form a capital L.

The University library will be 6 stories high and will have an aggregate floor area of 81,400 square meters. There will be a large atrium, safeguarded with a sprinkler system, as in other modern libraries. There will also be a system of rolling fire doors for isolating the atrium in a fire emergency. These are of interlocking steel slats and are wound on an overhead roll when not in use. In a fire they can close automatically on the action of a fusible link, or they can be closed manually. This was an extra fire defense demanded by the extraordinary volumes of space to be protected, and the potential very heavy fuel loading of these spaces. Conventional systems are planned for emergency lighting, smoke detection and smoke extraction, and Halon 1301 will be installed to protect rare collections.

Another fine library near Riyadh is that of the King Abdulaziz[9] Military academy, a 4-year academic and engineering center similar to West Point or the Air Force Academy in size and function. Here also the library is to be located at the very nerve center of the campus, occupying about a third of the combination library and administration building. A grand archway opening through this building will serve as the principal entrance to the Academy, where distinguished guests will be received.

The library will not only serve the mission of the Academy in education and research, but will be a depository for special collections of historical importance and archival records. Like the nearby University of Riyad library, it will have the most modern concept of fire protection for the books, including detection and suppression systems that will operate automatically and at the same time summon the Academy's own fire department.

Buildings Without Windows

A number of modern library buildings are largely without windows, and more will very likely be built, since this construction has obvious benefits for a library. The control of air pollution, temperatures, humidity and ultraviolet light are all simpler with a windowless building. So is defense against thieves and vandals. And there is a very great increase in useful wall space.

However, windowless construction drastically complicates the business of putting out a fire in the building. Unless special provision has been made for it, removal of smoke is almost impossible. Lights can fail and leave the building interior pitch dark at high noon. Worst of all, for lack of any way to vent it, the fire zone becomes superheated. Because of these problems, some codes call for emergency lighting and automatic suppression systems in windowless areas of buildings and in basements of all but very small buildings.

The problem was demonstrated in the spring of 1969 at Indiana University, when the second incendiary fire of the year struck an old library building. To vent the fire in the basement and introduce a conveyor to bring out smoldering debris the fire department had to drill with air hammers through a 36-inch limestone bearing wall. It was a 14-hour operation. Planners would do well to have professional advice on the design of fire protection for basements and windowless areas.

New Protection for Old Libraries

Attention to fire protection is probably more important in an old library than in any of recent construction. There must be hundreds of libraries built between 1850 and 1950, for example, still in daily service and storing important collections. Yet many buildings of that period, and even later buildings, are without fire divisions of any value, and have various other structural deficiencies. Many of them have open style book stacks in a free-standing structure of unprotected steel,[10] ready to soften and collapse in a fire. Most have no automatic systems for detection or suppression of fire. There is a real need for reexamination of any such library to see what can be done to improve its defense against fire.

A number of libraries in the University of California system have been vastly improved over the last few years. South Hall, more than 100 years old, houses the Library School on the Berkeley campus, and has its own distinguished library; it is completely protected with automatic sprinklers. The East Asiatic Library in Durant Hall has sprinklers in the public areas and the stacks, while rare books and items of special value are protected with a zoned Halon 1301 system.

In the same area are the Doe and Bancroft libraries, holding the largest concentration of books on the campus, occupying adjoining buildings and essentially forming a single fire risk. They have recently been fitted with a complex of automatic sprinkler systems at a cost well over $1,000,000. The cost was hardly significant in relation to the values in the collections protected. The improvement virtually removed the possibility of a shock loss from fire in this important property, and greatly stabilized the risk management program, in that university properties are essentially self-insured.

Sprinkler Systems—3 Types

The Doe and Bancroft Libraries are equipped with three types of water suppression systems. Conventional wet pipe systems are installed in public areas, offices and service areas. Book storage areas generally have a pre-action system; still others have that system refined with the use of the stop-and-go head. Early warning systems are on guard with smoke detection sensors, and Halon 1301 systems are planned for protection of collections of rare books and manuscripts.

Boalt Hall's law library at Berkeley has sprinkler protection. So does the Inter-Campus Lending Facility(North), a depository for more than a million books in a single huge room at Richmond, California. Other departmental libraries at Berkeley and on other campuses of the University have sprinkler protection in basement areas, detection systems without exception, standpipe and hose and first-aid fire equipment. The larger and more important libraries in the system are scheduled for automatic suppression systems of appropriate kinds.

Rutgers, the State University of New Jersey, several years ago provided a pre-action sprinkler system for their main library. The undergraduate library at Yale is sprinkler-protected. The McKeldin Library at the University of Maryland recently installed sprinklers after it was determined by insurance underwriters to be the site of the greatest concentration of values of any building in the state, and virtually uninsurable without an automatic suppression system.

Indiana University's very large modern library at Bloomington has an automatic detection system that has more than once operated to report incipient fires and harrassment attempts. Even better protected is the Lilly Library and Museum, also on the Bloomington campus in a 12-story building where the fire protection was improved long after the original construction. A complete automatic sprinkler system now protects the entire building, except that a fine Gutenberg Bible and other rare items are protected with a Halon 1301 arrangement. Both systems in the Lilly Building are operated with cross-zoned detection systems, which require impulses from two well-separated circuits to trigger the system, preventing premature action. The delay permits intelligent action to abort a nuisance fire, such as a wastebasket fire from a carelessly discarded cigarette, without activating a suppression system.

Life Safety, A Primary Responsibility

It is possible in the planning of fire protection and security of library collections to fail to provide adequately for the safety of people. In working to thwart thieves and vandals, for example, we must be careful not to put chains and padlocks on required exit doors. There is bound to be a better way. There is an electrically operated system for releasing door latches in an emergency, and this is accepted in some jurisdictions.

The layout or plan of any large library, particularly in the stacks, should be scrutinized for the safety of persons during an emergency. There are three minimum safety essentials for such a building: (a) Exit routes in the stacks should be clearly indicated with oversized arrows and color stripes on the floor of passageways; (b) Emergency lighting should be provided; battery powered, independent, self-charging units have proved satisfactory; (c) There should be a public address system. The code may demand a horn or a bell, but these are sometimes only confusing in the emergency,[11] whereas a P/A system permits intelligent information to be disseminated, as well as necessary routine announcements that may be necessary.

The breakdown of safety systems in a new addition to a Boston library in the summer of 1977 apparently led to the death of an elderly woman. In trying to leave the library she went through two doors marked "Emergency Exit Only," one of which was supposed to sound an alarm. She was then trapped in a passageway terminating on an exit door "fortified with granite" and weighing 800 pounds. She was found dead in the passageway 22 days later. *Library Journal*[12] commented "The incident is of significance to library security planners, who are reminded that the sophisticated security systems they're developing to protect libraries from today's thieves and vandals could hurt the ordinary library patron."

Any library officer may compare his exit facilities with the rules of the *Life Safety Code* of the National Fire Protection Association, NFPA #101 in the National Fire Codes. In this standard, in widespread use and adopted by reference in many cities and towns, a descriptive passage called Chapter 2, *Fundamental Requirements,* lays down the basic rules for safe exit facilities (p134).

John Sharry, staff specialist in life safety for the NFPA, points out that even compliance with the *Life Safety Code* may not always provide safety.[13] The user must not become so wrapped up in the mechanics and verbiage of the code that he loses sight of the spirit of it; common sense must rule. Sharry notes that the Code does not yet adequately provide for the handicapped, who may need wheelchair ramps, better designed steps, handrails, and other things. Rules for such improvements are spelled out in the standard ANSI A117.1 *Specifications for Making Buildings and Facilities Accessible to and Usable by the Physically Handicapped* (1961, revised 1971).

A new facility built expressly for the handicapped is the new Illinois Regional Library for the Blind on Roosevelt Road in Chicago.[14,15] It is part of the Chicago Library System and also operational headquarters for the Library of Congress, Division for the Blind and Physically Handicapped. A similar library has been built at Montgomery, Alabama. An earlier library restyled for the handicapped is that of the Human Resources School of Albertson, Long Island, New York.[16]

Noel Savage of *Library Journal* finds a growing awareness in libraries of the need for liability insurance protection, primarily for two reasons. One is the latter-day trend toward absolute liability, and the belief that almost any personal injury sustained on the premises of another can be the basis of a suit for damages. This belief is nourished by the enthusiasm juries seem to have for making grandiose awards.

Another reason is the need for dealing with the "problem patron" and for dealing skillfully, protecting other patrons and the library's property without incurring litigation for false arrest or some other charge. The increased need for defense against theft, vandalism, arson and crimes against persons demands training of staff to handle crime situations. To obtain resources for such training and to protect the library from litigation and financial loss in meeting these situations there is no substitute for an adequate liability insurance program.

Water Damage and Restoration

Water damage continues to occur in libraries even without fires. On a Sunday evening in July, 1975 a water line broke in an elevator shaft in Powell Library at the University of California at Los Angeles. Water trickled across ceilings and along book shelves, and quite a number of books got wet. Happily the water created a short circuit in fire alarm wiring, and help was at once on the way. Fire department pumps removed several inches of water from the lowest level. Some items could be air dried successfully at the library; others had to be sent to the vacuum chambers. Slightly damaged were some items from the Orsini collection, part of which dates from 998 A.D. About $22,000 was recovered in payment of insurance claims on books and equipment.

At Case-Western University in Cleveland a cloudburst and flooding produced a massive salvage problem on August 24, 1975. After four hours of downpour the water stood 8 ft deep outside the library and 43 inches deep inside. Soaked and muddy were 40,000 books, 50,000 maps and periodicals. Conservator Wilman Spawn of Philadelphia, placed in charge of the salvage project, declared the damage the worst he had ever seen and predicted it would take 12 months to repair.

Seven large freezer trucks were loaded with 40,000 pounds of wet papers and sent off to the vacuum chambers at McDonnell-Douglas, St. Louis. The cost of repairs to the building and furnishings was just under $30,000. The cost to restore the collections came to $540,146, including cleaning, rebinding and replacement of items that were beyond repair. Shipping cost $130,000 and rebinding $25,000.[17] Case-Western fared better than many other libraries struck by disaster, in that their insurance program was reviewed annually and brought up to current value. A "valuable papers" policy at $25 per volume made the costly restoration project possible.

Stanford University's J. Henry Meyer undergraduate library sustained a major water damage casualty in the early morning hours on November 4, 1978. Openings had been made in the building wall for pipe connections to the new Cecil H. Green library. A water main outside the building failed during the night and water poured in, flooding the lowest level in the Meyer library 12 inches deep. About 37,000 books were affected, including 3000 miniature books, some dating back into the 17th century; many books were designated "irreplaceable" and the total valuation was over $1,000,000.

Wet books were moved immediately to a San Jose cold storage facility in accordance with Stanford's library disaster plan, and preparations were made for freeze-drying and restoration. Construction operations in and adjacent to libraries have been responsible for fire disasters in libraries* more than once; this incident demonstrates the need for vigilance against water damage and flooding during such operations.

Most libraries are aware of the possibility of instant disaster, and many have made preparations for meeting an emergency. A recommendation to University

* p. 43, *Fire Prevention,* item 4

of California libraries proposed separate organizations to prevent disaster and to mitigate the severity of loss, a Disaster Prevention Team (DPT) and a Disaster Action Team (DAT) in each major library.

Organization for Salvage

Disaster planning commences with the systematic listing of resources, starting with the Library of Congress, then regional or local conservator talent, then local sources of help, if needed, for moving, quick-freezing, cleaning, deodorizing and dehydrating wet and smoky books. The need for strong and expert direction from a qualified conservator is paramount. Peter Waters' manual, available from the Library of Congress, is everywhere acknowledged as the Good Book of the restoration process.

In New England a formidable organization stands ready to meet library fire and water emergencies in any of the six New England states, plus New York.[20] This is the New England Document Conservation Center at Andover, Massachusetts, under the leadership of Director Emeritus George M. Cunha. The Center is governed by the New England Library Board, composed of state librarians cooperating under an agreement called the New England Interstate Library Compact.

When a disaster strikes, emergency teams are dispatched from the Andover headquarters post haste, like firemen on their way to a fire. They attack the emergency using Library Board funds for the initial effort, usually 24 hours, after which the local institution is able to carry on by itself.

During the initial period of management by NEDCC the damage is evaluated, salvage materials assembled, volunteer workers instructed in salvage techniques, and priorities for action are established.

Wheaton College Incident

This was the organization that responded when Wheaton College at Norton, Massachusetts sustained a water damage casualty January 25, 1976 in the Fine Arts Library. A window had inadvertently been left open in a studio on the floor above, and a water pipe froze and burst when the temperature dropped to –30°F. during the night. When the window was closed next day and the temperature rose to normal, water poured from the broken pipe and found its way into the library directly below.

When the librarian learned of the damage a call went at once to the Andover center. Within a short time George Cunha and his staff were on the scene at Wheaton. A massive air drying operation was set up under the direction of the NEDCC artisans, with Wheaton library staff and faculty volunteers doing much of the work. Electric fans were borrowed from townspeople. The air drying technique was feasible because most of the materials were damp, not sodden. Only 24 hours after the damage was done, all salvageable materials were drying. Ten days later the library was back in full operation. The loss was figured at $17,000, including replacement of the items destroyed.

It seems likely that regional conservation agencies like the New England Center will be set up in other areas. Certainly there is wisdom in the development of a strike force capability to meet library disasters with special skills and special equipment in any place where such services would be available to a large number of libraries.

Cost of Protection

The list of museums and libraries protected from fire with Halon 1301 continues to grow. Although it is relatively expensive, its special qualities in the protection of highly valued materials appear to justify the cost. When Harvard's Countway Medical Library was first provided with a Halon system for $40,000 a few years ago the value of the library was about $20,000,000. The initial cost was therefore roughly 2 percent,[21] and every year even this modest investment gets smaller in terms of annual cost of protection. Any expenditure for this system (or any other reliable system of fire protection) should be looked at in this light, amortizing the cost over the life of the system.

A major installation of Halon 1301 is that in the Nathan Marsh Pusey Library at Harvard University.[22] Pusey Library was built quite literally in the shadows of Widener and Houghton Libraries, and largely underground, so as to obtain another library and still preserve the character of the Harvard Yard. It has a total flooding system, funded by a generous donor, which directs the extinguishant selectively to any room where it is needed. Materials protected in this way include the letters of Amy Lowell and Ralph Waldo Emerson, memorabilia of Theodore Roosevelt, the Harvard Map and Theatre collections, and the University Archives.

The number of installations of Halon 1301 during the last ten years in the United States alone is estimated at more than 10,000. A few of these are in rare books. Unfortunately many fine collections are still unprotected by any suppression system other than the response of a fire brigade or fire department, which may or may not be called in time to remove rare books before taking care of the fire.

The use of Halon 1301 in portable appliances has been largely overlooked. The NFPA Handbook[23] shows it as 2½ times more effective than carbon dioxide "on a weight-of-agent basis," and effective against Class A fires (wood, paper, etc.) mainly because of cooling effect. It has also "the ability to penetrate obscured hazards in a way that dry chemicals cannot." It is the least toxic of gas extinguishants, and the only one sanctioned by NFPA for use in occupied spaces. To obtain its various advantages the Beinecke Library of Rare Books at Yale is considering converting from CO_2 to Halon 1301 protection.

Mobile Storage Equipment

New library equipment may introduce another fire protection problem. This was predicted in the 1975 edition of NFPA No. 910, *Protection of Library Collections,* in this passage: "The practice of mounting book stack ranges on

NATIONAL FIRE PROTECTION ASSOCIATION "DECISION TREE"

(NOTE: Figure represents only a segment of entire decision-tree schematic)

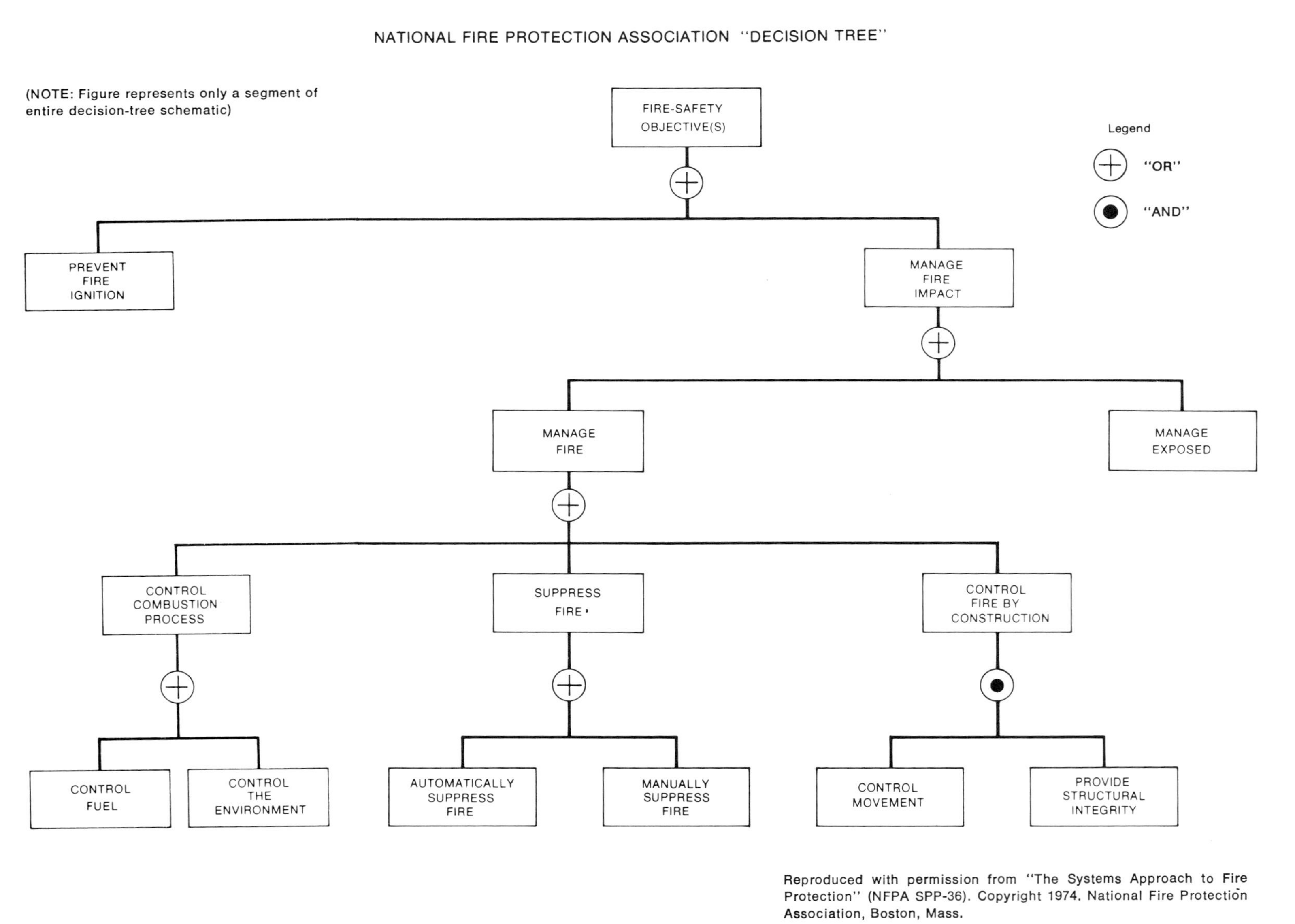

tracks is now appearing in library construction and renovations as an application of modern warehousing technology. This practice can result in fire durations in excess of eight hours—more than enough to challenge the strongest code-prescribed fire barriers and construction."

Tests performed by Factory Mutuals Research in January, 1978 for the General Services Administration[24] investigated the efficiency of automatic sprinklers in controlling fire in arrangements of compact storage. They indicated that conventional spacing of sprinkler heads might not provide adequate protection for this type of storage, particularly for plastic tapes and materials other than paper. These tests and their findings will have an impact on the planning of large installations of mobile book stack ranges in libraries. Plans for such an installation should be reviewed by a fire protection consultant to determine whether a special modification of any kind might be advisable to insure adequate and efficient fire protection.

There is another hazard with mobile storage systems, and particularly with motorized equipment; this is the hazard of being trapped and crushed between moving and stationary elements.[25] A safety engineering study should always be made to define the work hazards and set up safe practice routines. It is possible now also to obtain equipment in which the danger is controlled through electrical contacts and interlocks. In such a system there is a safety floor that immobilizes the equipment when a worker is standing in a position of possible danger, and a pressure sensitive device that stops it if he reaches into a danger point. Another arrangement will stop the equipment if a single book has been inadvertently left in its path and subject to damage.

For Want of a Nail

The tragedy of having a priceless collection inadequately protected against fire is tolled from time to time in the wake of disaster. On July 8, 1978 fire in the night destroyed the Museum of Modern Art in Rio de Janeiro. In less than 30 minutes 950 works of art representing some of the greatest painters of the 20th century were reduced to ashes.[26] The financial director of the museum said it would cost more than $7,500,000 to restore the concrete and glass building, "including the installation of an adequate sprinkler system."

The total loss in the fire was estimated at $15,000,000, which, even if it could be recovered, would not replace the lost paintings of Miro, Van Gogh, Picasso, Rivera and many others. It is not unlikely that the cost of an automatic sprinkler system for the building would have been well under one percent of the values protected, or even one percent of the value of the building without the exhibits.

The question to be asked once again after such an event is this one: "Why is it that we never can find money to provide for adequate protection of our most priceless resources, but after the disaster strikes we can always find money for rebuilding?" Surely responsible stewardship of treasured art and literary collections demands a superior level of protection. No museum or library of any stature can afford to rely upon a local(in-house) detection system, or watch services or watchmen, or proximity to a fire station, or a building of even the most modern fire resistive construction, none of which truly provides adequate protection.

REFERENCES

1. FIRE PROTECTION HANDBOOK, 14th Edition, National Fire Protection Association, Boston, 1976:6–5.
2. Wellesley, Massachusetts.
3. Downs, Barry V., "A Richness of Form, Space and Books," CANADIAN ARCHITECT, January, 1978:25–29.
4. "Metropolitan Toronto Library," CANADIAN ARCHITECT, January 1978:20.
5. Brook, Philip R., "Library Design for Today's User," CANADIAN ARCHITECT, January, 1978:30–35.
6. Hoyt, Charles King, "Relating Common Solutions: Two Libraries by Mitchell-Giurgola," ARCHITECTURAL RECORD, vol. 162, no. 2, August, 1977:93–98.
7. "The Public Records Office, Kew," FIRE PREVENTION, no. 125, June, 1978:16–19.
8. "Saudi Jobs," ENGINEERING NEWS-RECORD, November 11, 1976.
9. Also 'Abd al-Aziz.
10. Johnson, Edward M., Ed., "Protecting the Library and Its Resources, A Guide to Physical Protection and Insurance," Chicago, American Library Association, 1963:27–29.
11. LIBRARY JOURNAL, vol. 100, no. 8, April 15, 1975:712.
12. LIBRARY JOURNAL, vol. 102, no. 18, October 15, 1977:2106.
13. Sharry, John, "NFPA 101, Fundamental Requirements," FIRE JOURNAL, vol. 70, no. 2, March, 1976:73.
14. Goldberger, Paul, "Library for the Blind an Architectural Triumph," NEW YORK TIMES, August 9, 1978:C18.
15. LIBRARY JOURNAL, vol. 103, no. 14, August, 1978:1455.
16. Velleman, Ruth E., "Library Adaptations for the Handicapped," SCHOOL LIBRARY JOURNAL, vol. 99, no. 18, October 15, 1974:2713.
17. Correspondence from John Wesley Williams.
18. LIBRARY JOURNAL, vol. 101, no. 2, January 15, 1976:299.
19. LIBRARY JOURNAL, vol. 102, no. 8, April 15, 1977:858.
20. Correspondence from George Martin Cunha.
21. Welch, Claude E., M.D., "Fire in Libraries," NEW ENGLAND JOURNAL OF MEDICINE, vol. 287, no. 17, October 26, 1972.
22. ARCHITECTURAL RECORD, vol. 160, no. 4, September, 1976:97–102.
23. FIRE PROTECTION HANDBOOK, 14th Edition, National Fire Protection Association, Boston, 1976:13–23; Table 13–4C.
24. Chicarello, Peter J., et al., "Fire Tests in Mobile Storage Systems for Archival Storage," Technical Report No. K.I. 3A 3N4.RR, Factory Mutual Research, Norwood, Massachusetts, June, 1978.
25. LIBRARY JOURNAL, vol. 101, no. 2, January 15, 1976:298.
26. SAN FRANCISCO CHRONICLE, July 9, 1978.

Appendixes

FIRE PROTECTION AT THE NATIONAL ARCHIVES BUILDING

Washington, D.C.

LEO H. SWAYNE
Accident and Fire Prevention Branch
General Services Administration, Region 3

The National Archives Building, officially opened in November 1935, is one of the most imposing and monumental structures in Washington, D.C. Designed by John Russell Pope, who was architect for many of the outstanding edifices in the nation's capital, it serves as a special repository for the federal government's rarest and most valuable documents.

The design concept is a building-within-a-building. The outer structure, with its imposing display of columns and porticoes, forms a majestic enclosure for the inner core, which is the area specifically designed for 10 million cubic feet of storage. The space is arranged in 21 tiers, each approximately 500,000 cubic feet in size. Individual storage rooms or stacks range from 20,000 to 92,000 cubic feet. Special fire-resistive vaults and safes are included to house certain original and vital documents.

The original fire safety concept, when the building was constructed, was to house the records in noncombustible containers stored in compartments or stacks of fire-resistive construction that conformed with the best fire and life safety features in building design available at that time. The records storage areas were protected with a pneumatic-tube type of automatic fire detection system.

These measures proved adequate until the volume of records requiring storage exceeded the available space. To provide the additional space needed, the use of noncombustible containers was discontinued; instead, cardboard boxes holding loose papers and file folders were used and were stored on open metal shelving.

The photos in this article were provided by the National Archives.

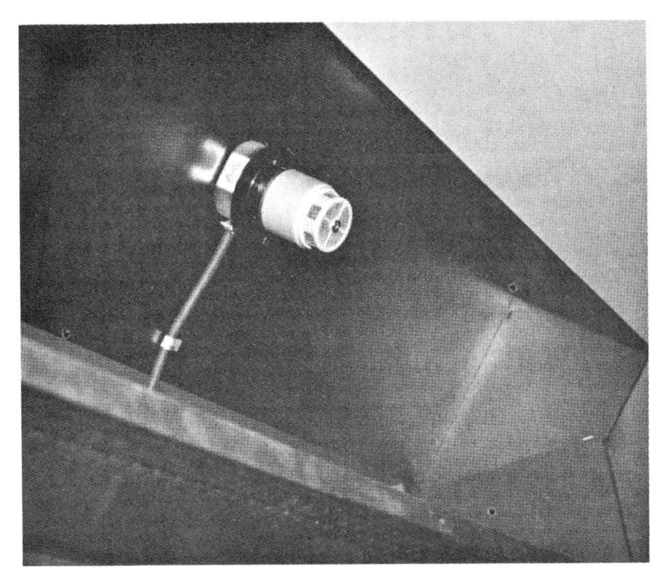

Ion chamber fire detector installed in a cabinet used to display valuable documents. These detectors are part of the overall fire detection system.

Their use required substantial reliance on controlling the possible sources of ignition and early detection of any fires that might occur. This was achieved by limiting persons admitted to the records storage areas, strict control of ignition sources, frequent patrolling by building guards, and by maintaining a high level of reliability in the automatic fire detection system.

Over a period of 20 years, the mass of records and the number and activities of employees and visitors in the building increased substantially. A few minor fires occurred, but only minor or no losses resulted. The occurrence of a "highly improbable" fire on February 7, 1965, in an area used to display original historical documents (including the Declaration of Independence), brought about recognition of the need for a more comprehensive system in the building to ensure against damage and/or loss from fire of the irreplaceable records and artifacts stored there.

Spurred by this fire, a study was immediately initiated to determine the best methods of achieving the desired level of fire protection. The study considered guidelines provided in the then-current (1963) NFPA Standard No. 232, *Protection of Records,* along with applicable criteria and guidelines of the General Services Administration. It also considered recognized fire protection engineering and fire control concepts in developing its recommendations.

The primary recommendations of the study called for: (1) installing early warning fire detection and automatic fire suppression systems in all records storage areas; (2) modifying the existing air conditioning/ventilating systems to permit controlled exhausting of smoke and heat; (3) removing all combustible interior finish and other decorative features or furnishings in records storage areas; and (4) developing a coordinated first-aid fire-fighting response to all fire, fire detection, and automatic sprinkler alarms. Because of the physical arrangements of the storage areas and equipment and the particular "shrine"-like nature of the records, a substantial engineering effort was required to develop a design that would integrate all of the desired fire protection features in the most effective and feasible manner.

Secondary recommendations were also made for updating and modifying existing fire protection systems to increase still further the overall protection of both the records and the building. The recommendations were implemented and work was started in early 1966. As the project proceeded, the most recent developments in fire detection, fire suppression and control systems, and alarm response procedures were constantly reviewed and additional desired features were included. The total system, completed in May 1974, represents the most advanced design concepts for safeguarding vital and valuable archival records from loss or damage by fire.

Typical arrangement of the automatic sprinklers and early warning-type fire detectors in records storage areas.

The nerve center of the records protection system is the alarm system console located in a special room and constantly manned by a Federal Protection Officer.

The total system includes the following features:

(a) A manual presignal fire/evacuation alarm system, with pull stations and evacuation bells, was installed throughout the building. In addition to registering the identity and location of the pull station actuated, an audible alarm is sounded in special locations, including the building guard office, key building maintenance personnel shops, and in other locations where personnel have been assigned specific duty in response to a fire alarm. The alarm is automatically transmitted to GSA's control center for transmission to the District of Columbia Fire Department.

(b) Ion chamber fire detectors were installed throughout all records storage areas and in all other high fuel load or high-value areas. Alarms from the fire detectors are arranged to register the location of the detector in alarm condition and to cause the same alarm transmission and response procedure as described for a manual fire alarm.

(c) Automatic sprinkler systems hydraulically designed to meet particular layout and discharge rates were installed in all records storage areas, maintenance shop, and other high fuel load areas throughout the building. All sprinkler waterflow alarms register and are automatically retransmitted and given the same response as a manual fire alarm.

(d) the ventilating/air-conditioning system has the capability to exhaust heat or smoke selectively from the individual records storage areas.

(e) A Class III standpipe and hose system, with hose stations having both 2½-inch and 1½-inch hose connections, is located at all floor landings in each of the four interior stairways to be used principally by the Fire Department.[1] Stored pressure water-type fire extinguishers are also located in all hose cabinets and at other strategic locations in the records storage areas for first-aid fire control.

(f) Procedures that call for the response of first-aid fire-fighting teams to all manual fire alarms, fire detector alarms, and automatic sprinkler alarms have been established.

(g) Combustible furnishings, equipment, and interior finishings have been minimized in records storage areas.

(h) Ion chamber fire detectors connected to the building fire detector system were installed in special storage vaults and document display cases.

While this $1.5 million project has resulted in an overall fire protection system directed principally toward the prevention of property loss, it has also improved the life safety features inherent in the initial design of the building to provide additional assurance for the safety of both the occupants and the hundreds of daily visitors to this monumental building. △

[1] Class III Service, as defined in *Standard for the Installation of Standpipe and Hose Systems*, NFPA No. 14–1973, is one that is capable of furnishing the effective fire streams required during the more advanced stages of fire on the inside of buildings, as well as providing a ready means for the control of fires by the occupants of the building.

Photo from Temple University News Bureau by William M. Smith

Volunteers wrap water-soaked books from Temple University Law Library in Philadelphia to be placed in food freezer plants after fire in library. Bottom, thermal-vacuum space simulation chamber at General Electric Space Division at Valley Forge, Pa., was used to remove water from books.

Space-Age Drying Method Salvages Library Books

BY ROBERT BURNS
Staff Correspondent

A space-age drying process is salvaging thousands of books, journals, pamphlets and documents, some of them going back to Colonial times, which were water-soaked during a five-alarm fire at the Law Library at Temple University in July 1972.

About 60,000 soggy volumes were removed from the 400,000-title library soon after the fire and placed in commercial food freezer storage plants and they are now being put through a drying process at the General Electric Space Division at Valley Forge, Pa. There, the books and papers are placed in a 3000-cubic-foot thermal-vacuum chamber that primarily has been used for testing spacecraft, such as the Nimbus weather satellites and the Earth Resources Technology Satellite, under simulated space conditions.

The day after the fire, Temple turned over the salvage problem to the Insurance Company of North America, which got in touch with Peter Waters, restoration officer at the Library of Congress. Waters went to Philadelphia the following day and arranged for the immediate delivery of packing cartons, rolls of freezer paper, roll dispensers, sealing tape, thymol crystals, denatured alcohol and polyethylene sheeting.

Sprayed to inhibit mold

Compressed air sprayers and self-contained masks were borrowed from the Philadelphia Fire Department to spray the books with an alcohol and thymol mixture to inhibit mold growth. The fire department cooperated by siphoning all the water possible from the basement, where some books had been in 3 feet of water.

On the Saturday after the Tuesday fire, a crew of volunteers reported to the library in response to pleas over radio and TV stations. The books were passed out by two human chains to four tables near the building. A pile of cut freezer paper was placed on each table and the volunteer nearest each wet pile of books picked up the top book and placed it on the top sheet of freezer paper, causing it to adhere to the paper. The book was pushed along the line to the next worker, who turned it over for the first wrap, etc., until it was placed in a carton at the end of the table. When a carton was filled, it was placed into a freezer trailer truck.

Work continued until midnight, and was resumed again on Sunday morning. By 5 p.m. Sunday, it was felt that all the salvageable books had been removed from the library. Some 24,000 cubic feet of wet books had been removed.

First attempts fail

The books were trucked to storage freezers in Philadelphia and Allentown, Pa., and Vineland, N.J., where they remained while methods of drying out the books were tested. The initial methods were unsuccessful because of the varying types of paper, sizes of books, etc. Freeze-drying, the process used in producing instant coffee, left the pages wrinkled.

The search for a better salvage method led to the GE Space Division. GE technicians constructed three four-wheeled carts, each having 14 racks with six shelves per rack. An electric heating pad was installed under each of the 252 shelves.

Frozen books are placed on the shelves and the loaded carts are raised to the chamber door by an elevator and then rolled into the cylindrical chamber. When loaded, the 42 racks hold from 3000 to 4000 volumes. When the three carts are in the chamber and the door is secured, most of the chamber's air pressure is removed. The heating pads then increase the shelf temperature and the water in the books boils off at 100°F instead of 212°F. The moisture condenses and freezes in large chunks of ice on the chamber walls. About 500 gallons of water are extracted from each load.

No wrinkled pages

The first batch of books was left in the chamber for six days and when removed, the warm books were entirely free of wrinkled pages and any musty odor. Even the dust associated with library books was missing.

The completion of the book salvage is expected to take about six months. Many of the books, it was reported, will need new bindings and covers.

George Reese of Temple's insurance office said the cost of restoration amounted to about $4 per volume as opposed to the $25 per volume cost of replacement. However, he was quick to point out that many of the volumes salvaged could not have been replaced at any cost.

According to INA, this is the largest book salvage job ever undertaken in the United States, and it is the first time aerospace facilities have ever been used for drying books. □ □

MILITARY PERSONNEL RECORDS CENTER FIRE

Overland, Missouri (July 12, 1973)

Part 2 - RECORDS RECOVERY (Salvage of Wet Papers)

This is the final installment of a two-installment report. The initial installment, which appeared in the May issue of FIRE JOURNAL, *contained a report of the fire that occurred July 12, 1973, at the Military Personnel Records Center; a discussion of the structural behavior of the reinforced concrete building during the fire; and a description of the emergency procedures that were established to salvage records and remove rubble. This installment deals with postfire recovery operations including salvage of wet records, data reconstruction, and fire protection requirements for records centers. It is in three sections, each by a different author. Evans Walker, National Archives and Records Service, reports on records recovery; Walter W. Stender, National Archives and Records Service, describes data reconstruction operations; and Harold E. Nelson, General Services Administration, discusses fire protection for records centers.*

RECORDS RECOVERY

Evans Walker, Archivist Specialist
National Archives and Records Service
Office of Federal Records Centers

While the fire at the Military Personnel Records Center in Overland, Missouri, was still raging, it became apparent that the recovery operation would be the most complex in archival history. Water damage would obviously be extensive. At the time, it was not known whether any records would survive the fire on the sixth floor; at first, this was believed highly unlikely. The full complexity of the recovery operation was not clearly defined until July 16, when it became possible for National Archives and Records Service personnel to enter the building, visit the fire-blackened sixth floor, and begin the task of damage assessment.

It was soon evident that the recovery effort would be more difficult and involved than had been expected. Records had survived on the sixth floor, some of them virtually unscathed, which needed immediate attention. The massive recovery effort began slowly as experts from the National Archives and the Library of Congress arrived at Overland to join the damage-assessment team.

One of the first problems the team faced was the fact that conditions were ripe for the growth of mold. The summer climate in Overland is noted for both high temperature and high humidity. Water was standing on every floor of the six-story building. There was little circulation of air within the building and the humidity, as monitored by hygrometers placed in strategic locations, was rising rapidly. Peter Waters, Restoration Officer of the Library of Congress, urged that thymol be sprayed in all of the records storage areas of the building except on the severely fire-damaged sixth floor. A local exterminating company was called in to undertake this spraying operation. In addition, it was recommended that borax be spread on the water- and sludge-covered floors to retard the growth of mold and slime. Eventually all floors and records storage areas received this treatment, although both operations were hampered by the fact that water continued to be poured on rekindled fires on the sixth floor for more than a week after the fire was officially extinguished on July 16. This water flowed downward through the building, causing problems and delays until late in July.

Between July 12 and July 23, when a demolition contract calling for removal of the sixth floor was awarded, efforts to remove as many water-damaged records as possible from the Center received top priority. Of the approximately 1.2 million cubic feet of records on floors one through five, only 10,068 cubic feet received water damage. Two methods of treatment were used to dry these records. The first was the use of a vacuum-drying process at the facilities of McDonnell Douglas Aircraft Corporation in St. Louis, and the second utilized facilities of the Civilian Personnel Records Center, also in St. Louis, where the wet records – in plastic milk container cases – were placed on open racks in an area with controlled humidity and temperature. Although the latter system worked quite satisfactorily, it was not capable of handling what eventually became a tremendous volume of wet material. Thus the vacuum-drying system at McDonnell Douglas became the primary system used in the drying process. Eventually, similar vacuum-drying facilities at a NASA installation in Ohio were also used.

Photo 1. Vacuum drying system at McDonnell Douglas Aircraft Corporation, showing plastic cartons of records.
NATIONAL ARCHIVES AND RECORDS SERVICE

Considerable confusion exists about the vacuum-drying process. It is not a freeze-drying process. Freeze-drying is particularly useful when it is necessary to hold materials in a frozen state in order to prevent mold growth and deterioration. On the other hand the vacuum-drying process, which utilizes space-simulation chambers, offers many advantages when large quantities of material are involved. The records were placed in plastic milk container cases, which were then stacked on 40-inch-by-40-inch pallets and loaded directly into the chamber. (See Photo 1.) The material was at ambient temperature. Air was then evacuated from the steam ejector to the point where the temperature in the chamber reached the freezing point. The chamber was then purged with hot, dry air until the wet material was warmed to 50° F.

The cycle was repeated. The number of cycles required depended on the initial condition of the material. This technique was developed when it became apparent that freeze-drying would take too long, and that there was no satisfactory way to supply the heat necessary for evaporation while the chamber was at low pressure. The effectiveness of this process is demonstrated by the fact that during a typical loading of one chamber, which held approximately 2,000 milk container cases, approximately eight pounds of water were removed from each case — with a total amount of nearly eight tons of water removed during each chamber loading.

The final step in records recovery takes place after the records have been dried and returned to the Military Personnel Records Center. As a result of the fire, a new computer-based index system was established which will include all the records recovered from the sixth floor and all records removed from other floors of the center because of water damage. The new registry system was named the "B" registry system. A new computer program has been prepared and as records come through the rehabilitation process, they are placed in folders, identifying data is key-punched, and computer-generated labels are affixed to the folders, which are then shelved and once again ready for reference use. Key punching is continuing on a two-shift basis, with a work force double the size of that ordinarily assigned to the task. After the work is completed, the Military Personnel Records Center will have a considerably expanded computer index capability which will materially benefit the reference operation of the Center, since all Army and Air Force inquiries can be checked against the computer index to determine the location of a record.

The completion of the indexing operation is still months ahead. However, all the records have been recovered from the sixth and lower floors; the temporary staging areas, which at one time included thirty tents and covered the Center's parking lots and lawns, are no longer in operation; and the drying facilities in St. Louis and in Ohio are no longer needed. We expect that all of the recovered records will be available for reference purposes by the time this report is published. Thus a major portion of the recovery operation will be ended.

LIBRARY FIRE EXPERIENCE, ENGLAND

August 6, 1969: Firemen were credited with "doing a wonderful job in saving the bulk of the book collection, worth many thousands of pounds," when Plymouth Proprietary Library burned. The blaze was one of several started by lightning during a storm. The building has 4 stories and basement, and filled with smoke as fire attacked partitions, walls and ceilings. Most of the books were saved by the good work of the firemen with waterproof sheets placed over bookshelves during the fire and afterward. The library was forced to close temporarily for the second time in 157 years. During the war years the entire stock of books was lost to enemy action, along with the building erected for the Library in 1812.[1]

March 10, 1971: One of the world's finest medical libraries was destroyed by an early morning fire at Radcliffe Infirmary in Oxford. It was on the fourth floor of a 4-story building, from which 125 seriously ill patients had to be removed to safety. A spokesman said "It was a disaster. Our library of more than 15,000 volumes was irreplaceable." The fire had started in the library, but there was no suggestion of arson. The amount of the loss was set at £ 250,000.[2]

March 13, 1971: The library at Goldsmiths' College, London, was struck by fire and lost 30 percent of its collections. A fine history collection was destroyed, also those on geography, the basic sciences, physical education, theatre and art. One of the two main wings of the building was lost, and about 25,000 books and bound periodicals.[3]

October 5, 1971: Fire in the night severely damaged the "nerve centre" of the Libraries of Enfield Borough, Essex. Ironically, the site of the facility was the old Southgate Fire Station in Palmers Green. Here was the distribution center for all new books going out to the borough's 16 libraries, and the repository for the Bibliographic section, containing records of the collections. These were in wooden file cabinets, which fortunately survived the fire. The book interchange service was seriously dislocated, and there was considerable damage to books and equipment. More than 30 firemen fought the blaze.[4]

March 14, 1973: A fire starting at the rear of the building at night destroyed Wishaw Public Library in Lanarkshire. Lost were almost 40,000 books and "some historical records which can never be replaced." Along with the 40-year-old library an extension costing £ 60,000, which was under construction, was also destroyed. The fire was seen for miles around.[5]

September 2, 1973: The Tothill Branch Library in Plymouth, a part of a community center complex, was destroyed by fire discovered at about 2 A.M. Firemen responded to eight calls on the 999 emergency system from neighbors of the library when the asbestos roof of the one-story building cracked. They were able to put the fire down in 25 minutes, but the books and records were all destroyed. The busy and popular library was moved to the community center while plans were being made for restoration of complete services.[6]

June 1, 1975: A fire discovered just after midnight burned out the staff and administrative wing of the Coventry Central Lending Library. The books were not badly damaged, but all library records were destroyed, and £ 40,000 worth of new computer equipment which was still being installed. Firemen kept the fire from reaching the books. There was speculation about the origin of the fire, which had apparently started in accumulated waste paper at the rear of the adjacent restaurant. Officials saw no evidence of arson.[7-8]

January 2,1978: Bridgewater Library at Bristol, built in 1918, was damaged by a fire that started in the roof of an old section of the building. A strong response by firefighters brought the blaze under control in 30 minutes, but there was smoke and water damage to thousands of books.[9] The fire was determined to be incendiary. Heaviest damage was incurred in the area of the issue desk and the reference library. The loss included about 2,500 records and cassettes and 30,000 books. Bridgewater Library is the largest library in Somerset County.

REFERENCES

1. WESTERN EVENING HERALD, Plymouth, Devon, August 9, 1969.
2. DAILY EXPRESS, London, March 11, 1971.
3. BOOKSELLER, March 27, 1971.
4. PALMERS GREEN AND SOUTHGATE GAZETTE, October 9, 1971.
5. Lanarkshire EVENING NEWS, March 15, 1973.
6. WESTERN EVENING HERALD, Plymouth, Devon, September 3, 1973.
7. COVENTRY EVENING TELEGRAPH, June 2, 1975.
8. BIRMINGHAM EVENING MAIL, June 2, 1975.
9. WESTERN DAILY PRESS, Bristol, January 3, 1978.

Code for Safety to Life from Fire in Buildings and Structures

NFPA 101 — 1976

CHAPTER 2 FUNDAMENTAL REQUIREMENTS

2–1 Every building or structure, new or old, designed for human occupancy shall be provided with exits sufficient to permit the prompt escape of occupants in case of fire or other emergency. The design of exits and other safeguards shall be such that reliance for safety to life in case of fire or other emergency will not depend solely on any single safeguard; additional safeguards shall be provided for life safety in case any single safeguard is ineffective due to some human or mechanical failure.

2–2 Every building or structure shall be so constructed, arranged, equipped, maintained and operated as to avoid undue danger to the lives and safety of its occupants from fire, smoke, fumes, or resulting panic during the period of time reasonably necessary for escape from the building or structure in case of fire or other emergency.

2–3 Every building or structure shall be provided with exits of kinds, numbers, location and capacity appropriate to the individual building or structure, with due regard to the character of the occupancy, the number of persons exposed, the fire protection available, and the height and type of construction of the building or structure, to afford all occupants convenient facilities for escape.

2–4 In every building or structure, exits shall be so arranged and maintained as to provide free and unobstructed egress from all parts of the building or structure at all times when it is occupied. No lock or fastening shall be installed to prevent free escape from the inside of any building.

Exception: Locks shall be permitted in mental, detention, or corrective institutions where supervisory personnel are continually on duty and effective provisions are made to remove occupants in case of fire or other emergency.

2–5 Every exit shall be clearly visible or the route to reach it shall be conspicuously indicated in such a manner that every occupant of every building or structure who is physically and mentally capable will readily know the direction of escape from any point. Each path of escape, in its entirety, shall be so arranged or marked that the way to a place of safety is unmistakable. Any doorway or passageway not constituting an exit or way to reach an exit, but of such a character as to be subject to being mistaken for an exit, shall be so arranged or marked as to minimize its possible confusion with an exit and the resultant danger of persons endeavoring to escape from fire finding themselves trapped in a dead-end space, such as a cellar or storeroom, from which there is no other way out.

2–6 In every building or structure equipped for artificial illumination, adequate and reliable illumination shall be provided for all exit facilities.

2–7 In every building or structure of such size, arrangement, or occupancy that a fire may not itself provide adequate warning to occupants, fire alarm facilities shall be provided where necessary to warn occupants of the existence of fire so that they may escape or to facilitate the orderly conduct of fire exit drills.

2–8 Every building or structure, section, or area thereof of such size, occupancy, and arrangement that the reasonable safety of numbers of occupants may be endangered by the blocking of any single means of egress due to fire or smoke shall have at least two means of egress remote from each other, so arranged as to minimize any possibility that both may be blocked by any one fire or other emergency conditions.

2–9 Every vertical way of exit and other vertical opening between floors of a building shall be suitably enclosed or protected as necessary to afford reasonable safety to occupants while using exits, and to prevent spread of fire, smoke, or fumes through vertical openings from floor to floor before occupants have entered exits.

2–10 Compliance with this *Code* shall not be construed as eliminating or reducing the necessity for other provisions for safety of persons using a structure under normal occupancy conditions, nor shall any provision of the *Code* be construed as requiring or permitting any condition that may be hazardous under normal occupancy conditions.

V.

CHRONOLOGICAL SKETCH OF THE DESTRUCTION OF LIBRARIES BY FIRE IN ANCIENT AND MODERN TIMES, AND OF OTHER SEVERE LOSSES OF BOOKS AND MSS. BY FIRE OR WATER.

(See *The Destruction of Libraries by Fire, by* C. WALFORD, pp. 66-70.)

(*a*) FIRES IN LIBRARIES.

B.C. 48. *Alexandria.*—The larger library destroyed during the occupation of Julius Caesar. The following is a brief account of this library: Ptolemy Soter founded an academy or society of learned men, for the use of whom he made a collection of choice books, which under his successors grew to prodigious bulk. The museum and library were first in that part of the city called Brucheion; afterwards a supplementary library was established in the Serapeion. In the war which Julius Caesar waged against the inhabitants of Alexandria, some of the ships which he was obliged in self-preservation to set on fire, drifted to the shore, and communicated their flames to the adjoining houses, which, spreading into the Brucheion quarter, consumed the noble library which had been the work of so many kings. Seneca says the number of volumes at the time of the fire was 400,000; but Aulus Gellius says 700,000. The library of the Serapeion remained until A.D. 389.

A.D. 188. *Rome.*—Great part of the capitol, a *famous library,* and several contiguous buildings, were utterly destroyed by lightning. Eusebius says it consumed whole quarters of the city, and in them *several libraries.*

273. *Alexandria.*—The larger library again destroyed by fire—by Aurelian.

476. *Rome.*—The Theodosian Library, estimated to contain 120,000 volumes.

640. *Alexandria,* December 22.—The library was burned by command of Amru, after the capture of the city by the Saracens. This library is said at the time of its destruction to have contained 500,000 volumes. In it, too, Cleopatra had deposited 200,000 volumes of the Pergamean Library, which Mark Antony had presented to her.

730. *Constantinople.*—A considerable library destroyed by fire.

781. *Constantinople.*—A dreadful fire consumed the greater part of the city, with the Patriarch's Palace, in which were the comments of St. Chrysostom on the Scriptures, written with his own hand.

802-7. *Constantinople.*—During the usurpation of Basilicas a fire happened, which consumed the greater part of the city, with the library containing 120,000 vols., and (it is stated) the works of Homer written in golden characters on the great gut of a dragon, 120 ft. long!

11th century. *Egypt.*—The library of the caliphs, said to have contained 1,600,000 vols., was burned.

1318. *Bristol.*—A commission issued at this date by the Bishop of Winchester, ordered an inquiry concerning the records of the Gild of the Kalenders in this city; which gild had been originally founded before the Norman Conquest for the purpose of keeping old records of Bristow (Bristol) and elsewhere. It was reported that, "by reason of a fire that happened in the place or library that was in the said church of All Sainctes," many of the records had been lost.—Toulmin Smith, "English Gilds," 1870, p. 287.

1440. *Megaspilaeon.*—The library of the monastery of Megaspilaeon, on Mount Cyllene, was burned.

1600. *Megaspilaeon.*—The library, which had been re-established, was again burned.

1649. *Mainz.*—The library of the Augustines was burned.

1666. *London* (Great Fire of). It was stated on the authority of the first Earl of Manchester, in "Al Mondo," that this fire destroyed books to the value of £100,000.

1671. *Madrid.*—Part of the library of the Escurial was destroyed by fire.

1685. *Venice.*—The library of the Canons of St. Anthony was burned.

1697. *Stockholm.*—The Royal Library burned.

1728. *Copenhagen.*—The library was burned.

1731. *Cottonian MSS.*—A fire broke out on October 22, at Ashburnham House (Westminster), which greatly damaged the King's and the Cottonian Libraries (which had been placed there during the preceding year); and which, as we know, formed the nucleus of the British Museum Library. Out of 958 MS. vols., 97 were burned, and 105 were more or less charred, and many others injured. For nearly a century some of the most precious of the MSS. remained as the fire had left them; but in 1824 Mr. Forshall, the then keeper of the manuscripts in the British Museum (the Cottonian collection had been moved to the Museum in 1753), made a commencement towards their restoration, which his successor, Sir F. Madden, successfully continued. More than 300 important MSS. have been perfectly restored. W. B[lades]., in the "Printer's Register," June, 1879, said hereon: "Much skill was shown in the partial restoration of many of the books damaged; some of them being charred almost beyond recognition. They were carefully separated leaf by leaf, soaked in a chemical solution, and then pressed flat between sheets of transparent paper. A curious heap of scorched leaves previous to any treatment may be seen in a glass-case in the MS. department of the British Museum."

1731. *Brussels.*—This same year there was a fire at the Archduchess's palace, in which all the records of state, preserved in the Royal Chapel, were destroyed.

1748. *London,* March 25.—A fire broke out on Cornhill, and consumed 200 houses. The library and works of art of Gustavus Brander, F.R.S., F.S.A., were placed in great jeopardy by this calamity, but were mostly saved. It was supposed that Dr. Johnson witnessed this fire, and soon afterwards he wrote: "The conflagration of a city, with all its turmoil and concomitant distress, is one of the most dreadful spectacles which this world can afford to human eyes."

1750. *Newcastle-upon-Tyne,* August 28.—In this great fire the house and stock of Bryson, the celebrated bookseller, were destroyed.—Curwen's "History of Booksellers," p. 449.

1752. *Lincoln's Inn.*—A fire in the chambers of the Hon. Charles Yorke robbed the world of an invaluable collection of manuscripts and pamphlets, which had been gathered from various sources, with immense pains, by Lord Chancellor Somers.

1770. *London,* January 8.—Great fire in Paternoster Row; large stocks of books burned.

1777. *Bonn* (Germany), January 15.—The palace of the Electoral Prince of Cologne burned; many valuable books and works of art destroyed.

1780. *London,* June.—During the Lord George Gordon "no popery" riots, the house of Lord Mansfield, with untold manuscript treasures, was burned.

1794. *Paris.*—The library of the Abbey of S. Germain des Près was burned.

1805. *Neisse* (Prussia).—The library burned during the siege.

1812. *Moscow.*—In the burning of this city, amongst other treasures consumed was the library of the Aristotelian commentator, Joh. Th. Buhle.

1815. *London,* March.—The stock of Kelly, the great bookseller in Paternoster Row, burned. Curwen, in his "History of Booksellers" (p. 368), gives the following account of the circumstances: "It was about this time, in March, 1815, that he very nearly lost a moiety of his fortune through fire. Luckily, upon the outbreak of a fire in the neighbourhood a few days before, he had been alarmed, and had gone straightway to the office of the Phœnix Company, and paid a deposit on the insurance. Before the policy was made out, the *whole of his stock* was destroyed, but the Phœnix Company paid up without an hour's delay, and in return he never cancelled a single policy with them until this sum had been reimbursed."

1827. *Abo* (capital of Finland), Sept. 4.—The university, cathedral, and library of about 40,000 volumes destroyed.

1838. *Temple* (London).—Fire originated in Mr. Maule's (afterwards Mr. Justice Maule) chambers, and burned about twenty other sets of chambers. Several valuable libraries destroyed.

1845. Sept. 24.—Library of the Marischal College, *Aberdeen*—Dalgetty's "Alma Mater." The fire originated in the Latin class-room, and although the students worked at the removal of the books until they were half-smothered, the library was considerably damaged. The interest about this library is that, under its then title of the Aberdeen Institution (before the fusion of the Aberdeen Universities), it was entitled to a copy of all books entered at Stationers' Hall.

1860. *Bristol,* Feb. 14.—Premises and stock of Mr. Thomas Kerslake, an eminent bookseller, Park Street, Bristol, destroyed.

1860. *Dublin,* Nov. 11.—The Kildare Club House and large library burned.

1861. *London,* Sept. 4.—Fire broke out at Knight's, the tallowmelters, in Paternoster Row, and burnt out Messrs. Longman's and Co., destroying their valuable stock of old books.

1862. *London.*—The curious old library placed in a gallery in the Dutch church in Austin Friars, was nearly destroyed on the burning of the church this year. There were many fifteenth century books. The whole collection had been sadly neglected for many years, and was in a bad state of dust and dirt; to which I think, perhaps, it owed in some part its preservation. The water used at the fire converted the dust into a form of mud, or "muddy-pulp," as Mr. Blades has more graphically said. I will quote his description of what followed:—

"After all was over, the whole of the library (no portion of which could be legally given away) was *lent for ever* to the corporation of London. Scorched and sodden, the salvage came into the hands of Mr. Overall, the indefatigable librarian. He hung up the volumes that would bear it in a hired attic, over strings, like clothes to dry, and there, for weeks and weeks, were the stained, distorted volumes, often without covers, often in single leaves, carefully tended and dry-nursed. Washing, sizing, pressing, and binding effected wonders; and no one who to-day looks on the attractive little alcove in the Guildhall Library, labelled 'Bibliotheca Ecclesiae Londino Belgicae,' and sees the rows of handsomely lettered backs, could imagine that not long ago this, the most curious portion of the city's literary collection, was in a state when a five-pound note would have seemed more than full value for the lot."— "Printer's Register," June, 1879.

[An excellent catalogue of this library has just been prepared by Mr. Overall, 1879.]

1865. *London*, July 2.—The Offor Collection.—While this library was on sale at Sotheby's, and at the conclusion of the first day (the more valuable lots then sold being fortunately removed), a fire occurred in the adjoining house, and gaining possession of the sale-room, made a speedy end of the unique Bunyan and other rarities then on show. "I was allowed to see the ruins on the following day. . . . . It was curious to notice how the flames, burning off the backs of the books first, had then run up the backs of the shelves, and so attacked the fore-edge of the volumes standing upon them, leaving the majority with a perfectly untouched centre, while the whole surrounding paper was but a mass of black cinders."—"W. B[lades].," in "Printer's Register," June, 1879.

By this same fire there was also destroyed the Humboldt Library, consisting of about 17,000 volumes. I fear a good friend of our Association (Mr. Henry Stevens) was a heavy loser by this event. A £5,000 insurance expired at noon on the very day of the fire, and had not been renewed.

1865. *British Museum*, July 10.—The binders' room burned; several valuable MSS. destroyed.

1866. *Crystal Palace*, Dec. 30.—Library burned during destruction of north wing of Palace.

1870. *Strasburg.*—The magnificent library in this city was burned and otherwise destroyed by the shells of the German besieging army. "And over all the din of war the burning leaves of many a priceless volume floated on the heated air for several miles."—W. B[lades]., in "Printer's Register." I had an opportunity of seeing the remains of this once noted library.

1871. *Chicago.*—In addition to the many public and private libraries destroyed by the great conflagration in this city, there was one of especial interest, which had been founded by Mr. Yeager, then the editor of an insurance journal (the "Herald") published there. He had an almost unique collection of insurance newspapers.

1872. *Palace of the Escurial* (Spain), Oct. 2.—The grand library here was in great jeopardy, but was ultimately saved.

1873. *Manchester Athenæum*, Sept. 24.—Building partly burned, and 19,000 volumes.

1874. *Pantechnicon* (Pimlico).—It was understood that several libraries were stowed away in this building at the time of its destruction.

1875. *Advocates' Library* (Edinburgh), March 28.—Building on fire, and one of the largest and best libraries in the kingdom placed in jeopardy; but the fire was happily extinguished before much injury was done.

1876. *Messrs. Chick's Furniture Repository* (Paddington), June 2.—It was reported that several private libraries were included in the extensive destruction here.

1877. *St. John* (New Brunswick), July.—There were thirteen churches destroyed, with a corresponding number of parsonages, each of which contained a *minister's library.*

1877. *Philadelphia.*—The Mercantile Library burned: the insurance companies paid £8,400, and allowed the library to keep the damaged books. The building and furniture were also insured.

1878. *Madrid.*—Palace of the Duke of Villahermosa burned. Many fine examples of early and later art were consumed, together with the library,—the result of three centuries of collection,—which contained, in addition to many rare early editions of the classics and the fathers of the church, rare works of all kinds in choice sixteenth century bindings. It included the library belonging to the two Argensolas, Lupercio and Bartolomé. It was hoped there would be considerable salvage as to the books.—"Athenæum," 1878, p. 190.

1879. *Birmingham Free Library*, January 11.—The destruction of this library was regarded as one of the most serious losses of modern times. The library consisted of several sections: the reference portion had over 50,000 volumes. The famous Shakespeare Memorial Library contained about 7,000 volumes; and there was the Staunton Warwickshire Library, and the Cervantes Library, in addition.

The insurance offices settled the loss for £20,000: in the proportion of £8,000 for the building, and £12,000 for the books;—this it was estimated would leave a loss of about £12,000. Many voluntary donations of books have been made.

This disastrous fire has had the effect of drawing considerable attention to the unsafe condition of other public and private libraries. As an instance, the Mitchell Library in Glasgow was found to be surrounded with many "external risks." Steps were taken in view of removing it to a more safe location.—"Finance Chronicle," February 15, 1879.

1879. *Boston* (U.S.), Dec. 28.—A large fire did considerable damage to the stock of several booksellers and publishers, chiefly to that of Messrs. Houghton, Osgood, and Co., publishers; and Messrs. Rand, Avery, and Co., printers.

1879. *Clumber House* (Worksop).—This, the seat of the Duke of Newcastle, was found to be on fire; books in principal library removed, but much damaged by fire and water. Many books and records in other parts of mansion destroyed.

1879. *Irkutsk* (Siberia), August.—Extensive conflagration destroyed all the libraries in the town, viz. the Public Library, the private one of M. Vaghine (which contained the unpublished MSS. of Gedenstrom), and that of the Siberian branch of the Russian Geographical Society, which latter contained a great variety of works about Siberia, some of them being very rare; a great number of works and MSS. on Buddhism, numerous collections of publications of foreign scientific societies (European, Asiatic, and American), who exchanged their publications with the Siberian branch, and a large assortment of works on physical sciences and natural history. The destruction of this last-named library is regarded as a serious loss to men of science.

Note.—The foregoing is obviously an incomplete list; it must be regarded as a first effort only. Those who can add other details will confer a favour by so doing. It is obvious that the destruction of country mansions must cause the sacrifice of libraries, many of which are valuable, but are unknown to fame.

(*b*) DESTRUCTION OF PRINTING OFFICES, AND OF INDIVIDUAL BOOKS BY FIRE.

In addition to the preceding wholesale destruction of libraries as such, there are many instances on record of the destruction of individual works by fire. It is indeed probable that printing-houses hardly ever fall as sacrifices to the flames, without involving some loss to literature in this way.

1708.—The second volume of "An Ecclesiastical Parochial History of the Diocese of London, Herts, and Essex," by R. Newman (two volumes, small folio), is rendered very scarce by reason of almost the entire issue having been burned. I am not familiar with the circumstances.

1713. Jan. 29.—The printing office of Mr. Bowyer was totally destroyed by fire, together with "many considerable works."—Nichols, "Literary Anecdotes," vol. i., page 50.

1791.—Richard Warner's "Antiquitates Culinariae," curious tracts relating to the culinary affairs of the old English, with notes (4to. coloured plates), was nearly all destroyed by fire on the eve of publication.

1803. Feb. 2.—The printing house of Mr. S. Hamilton, in Falcon Court, Fleet Street, and other property destroyed. Many books in course of printing burned.

1808. Feb. 8.—The printing offices of Messrs. Nichols and Son, of Red Lion Court, Fleet Street, destroying (*inter alia*) the unsold numbers of the "Gentleman's Magazine" from 1783 to 1807; see also p. 153.

1821. Feb. 7.—The Caxton Printing Office, Liverpool, "the largest periodical printing office in the world," burned, including upwards of 3,000,000 of folio, 4to., and 8vo. numbers. Loss very considerable. Insured, £36,000.

1822. March 2.—Mr. Bagster's stock-in-trade in Paternoster Row burned.

1824. Aug.—The printing houses of Mr. Moyes and Mr. Wilson, in Greville Street, Hatton Garden, destroyed. Many books in course of printing were burned, including Corbaux's "Further Inquiry into the National Debt."

1830. August 11.—Fire on the premises of Mr. Adlard, printer, Bartholomew Close. The stock of the "Encyclopædia Londinensis" destroyed; was valued at £12,000.

1836. Feb. 18.—The Methodist bookstore, printing offices, &c., in New York, were burned. Stock valued at £50,000. Insurance, £5,000.

1837. March 20.—Fire at Mr. Spottiswoode's, "King's Printer." Warehouse containing printed sheets of "Lardner's Cyclopædia," Lord Byron's works, &c., burned. Loss, £20,000.

1852. June 10.—The establishment of Messrs. Clowes, printers, Duke Street, Lambeth, partly destroyed. A warehouse in which the fire originated, and which suffered the greatest destruction, contained (*inter alia*) the following: "Knight's Illustrated Bible," the "Sunday Book," the "Illustrated Shakespeare," the "Royal Catalogue of the Exhibition of all Nations," the "Church Catechism," "Readings for the Rail," and "Population Tables," being the results of the census of 1851.

1853. Nov. 10.—Messrs. Hooper's great printing and publishing works, in New York, burned. Numerous books in course of publication and stereotype plates destroyed. Loss, £275,000.

1861. Sept. 4.—At Messrs. Longman's fire in Paternoster Row at this date (*inter alia*) the new edition of "Tooke's History of Prices" was burned.

1878.—Messrs. Nelson's extensive printing and publishing works in Edinburgh were destroyed, including the stock of many well-known works.

1880. *London*, March 31.—Messrs. Dickens and Evans's printing works, New Street Square, Fleet Street, entirely destroyed, including various works just printed.

In this connexion I may note several "calamities to authors" which have occurred in the way of the burning of the materials which they were preparing for special works.

The first which occurs to me is that which happened to Sir Isaac Newton, by his pet dog upsetting a lighted taper, whereby were destroyed all the MS. notes concerning the experiments on colours and light upon which he (the master) had been engaged for several years. In true philosophic spirit he simply exclaimed, "O little dog Diamond! little dog Diamond! what hast thou done?"[1] Yet the loss affected his health and spirits: "He was not himself," says a contemporary writer, "for a whole month afterwards."

There is the well-known case of the late John Nichols, the antiquary. When he had about half printed his "Literary Anecdotes," the printing-office of his firm was (February 8, 1808) burned, and all his vast stores of yet unprinted notes and MSS. destroyed. He says, in the preface to his 2nd edition (1812): "In May, 1802, I once more began to print; and by slow degrees had got through nearly half the work, when my progress was suddenly retarded by a calamity which had well nigh disheartened me from again resuming the task either of editor or printer. *But on a serious conviction that despair was equally useless and criminal, I determined to begin my labour anew;* the fruits of which, such as they are, after being four years longer in the press, are again submitted to the public."

A more recent case is that of Admiral W. H. Smythe (author of "Ædes Hartwellianæ"), who, having had his library, and literary collections of a lifetime, including a good deal of MSS., destroyed by fire (about 1848?), set himself to work, in his old age, to "repair damages,"—such was his sailor-like expression; and he accomplished a considerable amount of good work afterwards.

(*c*) DESTRUCTION OF PUBLIC RECORDS, ETC.

Not only does fire destroy books, but it also destroys the materials from which many of our best books are made—I mean our national and our institutional and family records. It is in this light that our present Historical Manuscripts Commission is doing so much good; for, by recording what is now existing, and where, we shall have better means of knowing what in future shall be destroyed by individual fires; while the substance of the documents is frequently recorded in the Reports of the Commission.

Herbert, in the "Advertisement" to his "History of the Twelve Great Livery Companies of London," says (I. p. vii.): "The Fire of London, and other accidents, have left few of them (the City Companies) any records beyond the reign of Elizabeth." I suspect that in some of these cases the records may have been purposely destroyed during the reigns of Henry VIII. and his son Edward VI.

It appears, indeed, that the destruction of our records has not been, in many cases, the result of accident. Nash, in the preface to "Haue with you to Saffron Waldon, 1595," says that Polidore Virgile, in Henry VIII.'s time, burnt all the ancient records after he had finished his chronicle. I hope such an example has not been often followed; but there are at least some others on record.[2]

It is certain that nearly every public building that is burned leads to the destruction of some records, be their value great or small. The following are cases in point:—

1619. *Whitehall.* The fire at Whitehall in January of this year occasioned some loss and very serious confusion to the State Papers. Such as were saved were tossed into blankets and removed with much confusion as to their arrangement, from which they never fully recovered. Some of the more important had been removed just previous to the fire, owing, as Sir Thomas Wilson flatteringly asserted, to the "prophetical spirit" of the King, James I. *See* Edwards's "Libraries and Founders of Libraries," p. 187.

[1] This story, so far as the "little dog Diamond" is concerned, is of course apocryphal.—EDD.

[2] Frascator, one of the most eminent scholars of the sixteenth century, had written in Latin an elaborate history of Venice, which all who saw the manuscript praised in the highest terms. But the unfortunate author got up in the paroxysm of a fever, and committed the manuscript to the flames, and thus deprived scholars of a work "which need have feared no comparison with that of Livy." Longolius' conduct was still more extraordinary, especially as he could plead no such delirium as an excuse. He became so infatuated with the style of Cicero, that he not only determined to imitate his composition for the rest of his days, but to destroy everything he had written before he became acquainted with the writings of the great Roman stylist. Accordingly, he destroyed not only a valuable commentary on Pliny, but a considerable mass of manuscript matter as well; and, as he died shortly after his injudicious act, he left scarcely anything behind him.

1814. *Custom House* (London) burned. Many most important records destroyed.

1834. *Houses of Parliament* burned. Many records perished; but happily these had mostly been printed.

1838. *Royal Exchange.*—All the early records of the Royal Exchange Insurance Corporation were destroyed. They had been placed in vaults for safety.

1841. *Town Hall* (Derby).—All town records destroyed.

(*d*) DESTRUCTION OF BOOKS BY WATER.

It has always to be remembered that, next to fire, water is a great destroyer of books, and many a valuable library has been very seriously damaged by ill-advised attempts to save it; although, after the fire at the Dutch church in Austin Friars, referred to on page 150, we need not consider the damage from water thrown on to books to save them from burning as being beyond remedy. Even when books escape from a fire (as is often the case) in a parboiled condition, it is still only a question of patient but skilful treatment.

In this (the water) aspect of the case, I cannot help remarking upon the woeful catalogue which might be made of the literary treasures which have perished in the sea. May I not suggest this as an interesting subject of future inquiry, adding one or two incidents by way of a start?

15th century.—In the early part of this century a whole shipload of classical MSS., being removed from Constantinople by Guarino Veronese, went down. The owner survived; but his anxiety and grief were so great that his hair is said to have turned white in a few hours.

1600.—The magnificent collections of Vincentio Pinelli were on his death, this year, being transferred to Naples in three ships. One of these, however, fell into the hands of some pirates, who boarded it and flung most of the books and manuscripts into the sea, among them some very valuable antiquarian treatises; the rest of them were scattered up and down the neighbouring shore, and used, says Tiraboschi, who tells the tale, either for stuffing up the holes in boats, or to patch up broken windows.

1685.—John Dunton, the famous bookseller, sailing for Boston, U.S., lost a "venture of books," which, however, he only valued at £500.—"Life and Errors," p. 87.

1698.—A Dutch merchant, named Hudde, having made his fortune, resolved to devote the rest of his life to studying Chinese, and, being still a young man, sailed to China for the purpose of intimately acquainting himself with the details of Chinese government, as well as its history and literature. He was so fortunate as to have passed himself off for a Chinese, was raised to the dignity of a mandarin, and was thus enabled to master every detail of the difficult task he had undertaken. He had embodied the labour of thirty years in a prodigious mass of manuscripts, and on his return to Europe he was shipwrecked, and all his cherished treasures went to the bottom of the sea.

"Ibi omnis
Effusus labor."

1865.—Mr. Thomas Lidbitter, now an average stater in Bombay, lost, about this date, a considerable insurance library at sea, passing from Australia to Bombay.

(*e*) IMPERISHABLE RECORDS.

Final Note.—The Assyrian Empire went down in blood, and its palaces were destroyed by fire; but it had, either from wisdom or necessity, recorded its history in tablets of clay,—and these, which formed the treasures of its libraries, are preserved (more or less damaged, it is true) even to us. May we not wisely take a hint from this event, and record the great charters of our liberties and other important documents in a like imperishable form? Terra-cotta is wonderfully available for such purposes.

Bibliography

BOOKS

*"Approval Guide, Equipment, Materials, Services for the Conservation of Property," Norwood, Mass., Factory Mutual System, (annually).

Barlay, Stephen, "Fire, An International Report," Brattleboro, Vermont, Stephen Greene Press, 1969.

*"Best's Loss Control and Underwriting Manual," Morristown, N.J.: A. M. Best Co., 1968.

Bland, Richard E., "America Burning," Supt. of Documents, Washington, D.C.: National Commission on Fire Prevention and Control: 1973.

Bostwick, Arthur E., Ed., "The Library and Its Home," New York, H. W. Wilson: 1933.

Brannigan, Francis L., "Building Construction for the Fire Service," Boston, National Fire Protection Association, 1971.

Bryan, John L., "Automatic Sprinkler and Standpipe Systems," Boston, National Fire Protection Association, 1976.

Cunha, George Martin and Dorothy Grant Cunha, "Conservation of Library Materials, a Manual and Bibliography on the Care, Repair and Restoration of Library Materials," Metuchen, N.J., The Scarecrow Press, 1971.

Fire Record Bulletin FR 60-1, "Occupancy Fire Record, Libraries," Boston, National Fire Protection Association, 1961.

Jenkins, Joseph F., Ed., "Protecting Our Heritage, A Discourse on Fire Protection and Prevention in Historical Buildings and Landmarks," NFPA No. SPP-15, 2nd Edition, Boston, National Fire Protection Association, 1970.

Jensen, R., Ed., "Fire Protection for the Design Professional," Boston, Cahners Publishing Co., 1975.

Johnson, Edward M., Ed., "Protecting the Library and Its Resources, A Guide to Physical Protection and Insurance," Chicago, American Library Association, 1963.

Martin, John H., Ed., "The Corning Flood: Museum Under Water," Corning, N.Y., The Corning Museum of Glass, 1977.

McKinnon, Gordon, Ed., "Fire Protection Handbook," 14th Edition, Boston, National Fire Protection Association, 1976.

Metcalf, Keyes D., "Planning Academic and Research Library Buildings," New York, McGraw-Hill, 1965.

Moll, Kendall, "Arson, Vandalism and Violence: Law Enforcement Problems Affecting Fire Departments," U.S. Department of Justice, 1974.

Nash, P. and R. A. Young, "Automatic Sprinkler Systems for Fire Protection," London, Victor Green Publications, 1978.

**National Fire Codes:* Boston, National Fire Protection Association:

*NFPA #10, Installation, Use and Maintenance of Portable Fire Extinguishers, 1975.

*NFPA #12A, Halogenated Extinguishing Agent Systems-Halon 1301, 1973.
*NFPA #12B, Halogenated Agent Extinguishing Systems-Halon 1211, 1977.
*NFPA #13, Installation of Sprinkler Systems, 1975.
*NFPA #70, National Electrical Code, 1975.
*NFPA #101, Life Safety Code, 1976.
*NFPA #232, Protection of Records, 1975.
*NFPA #232AM, Archives and Records Centers, 1972.
*NFPA #910, Protection of Library Collections, 1975.
*NFPA #911, Protection of Museum Collections, 1974.
*NFPA Fire Protection Reference Directory, 1978.
NFPA #SPP-36, "The Systems Approach to Fire Protection," 1974.

National Survey on Library Security, Briarcliff Manor, N.Y., Burns Security Institute, 1973.

Rapoport, Roger and Laurence J. Kirshbaum, "Is the Library Burning?," New York, Random House, 1969.

Roseberry, Cecil R., "A History of the New York State Library," Albany, The State Education Department, 1970.

Singer, Dorothea M., "Insurance for Libraries, A Manual for Librarians," Chicago, American Library Association, 1946.

"People Care During a Fire Emergency, Psychological Aspects," Boston, Society of Fire Protection Engineers, 1975.

Trezza, Alphonse F., "Library Buildings: Innovation for Changing Needs," Chicago, American Library Association: 1972.

*Underwriters' Laboratories, Inc., Product Directories: "Fire Protection Equipment List; Accident, Automotive and Burglary Protection Equipment Lists"; Chicago, Underwriters' Laboratories, 1978.

Waters, Peter, "Procedures for Salvage of Water-Damaged Library Materials," Washington, Library of Congress, 1975.

Williams, John C., Ed., "Preservation of Papers and Textiles of Historic and Artistic Value," American Chemical Society, *Advances in Chemistry* series, Washington, D.C., 1977.

PAMPHLETS

American Insurance Association Occupancy Bulletin, "Libraries," New York, 1971.

Boehm, Hilda, "Disaster Prevention and Disaster Preparedness," University of California, Office of the Assistant Vice-President, Library Plans and Policies, Berkeley, Calif., 1978.

Cornell University Libraries, "Emergency Manual," C.U.L., Ithaca, N.Y., 1976.

Factory Mutual Engineering Corporation, "Arson," Norwood, Mass., 1977.

Factory Mutual Engineering Corporation, "Protection Against Incendiary Fires," Norwood, Mass., 1977.

Fire Protection Design Guide No. 6, "Equipment for Detection and Warning of Fire and Fighting Fire," in the series *Fire and the Architect,* Fire Protection Association, London EC4N 1TJ.

General Adjustment Bureau, "Was It Arson?" New York, N.Y., 1975.

* Publications of this organization are frequently revised; the current edition is usually preferred.

Harvard College Library, "Change and Continuity in the Harvard Yard—the Nathan Marsh Pusey Library," Cambridge, Mass., 1976.

Hellmuth, Obata and Kasselbaum, *"Libraries,"* St. Louis, Mo., 1975.

Matthai, Dr. Robert A., "Protection of Cultural Properties During Energy Emergencies," Arts/Energy Study and American Association of Museums, New York, N.Y., 1978.

Metropolitan Toronto Library Board, "Official Opening of the Metropolitan Toronto Library, November 2, 1977," Toronto, 1977.

Moody, Tom and others, "The Big Issues in Fire Safety, A Special Report from the National League of Cities," *Nation's Cities,* Mar., 1978.

Suchy, John T. and others, "Arson: America's Malignant Crime," U.S. Department of Commerce, Final Report of Leadership Seminars, Sept., 1976.

ARTICLES

Banks, Paul N., "Environmental Standards for Storage of Books and Manuscripts," *Library Journal,* Vol. 99, No. 3, Feb. 1, 1974:339–343.

Brook, Philip R., "Library Design for Today's User," *Canadian Architect,* Jan. 1978:30–35.

Brewer, Norval L., "Fire Destroys Aerospace Museum," *Fire Engineering,* Vol. 131, No. 6, June, 1978:24–25.

Burns, Roberts, "Space Age Drying Method Salvages Library Books," *Fire Engineering,* Vol. 126, No. 12, Dec. 1973:52.

Cohen, Wm., "Halon 1301, Library Fires and Post-Fire Procedures," *Library Security Newsletter,* May, 1975:5–7.

Corbett, Dennis F., "Halon 1301: A Fire Suppressant that Respects Rare Books," *Harvard Magazine,* Vol. 78, No. 9, May, 1976:12.

Cotton, P. E., "Fire Tests of Library Bookstacks," *NFPA Quarterly,* Vol. 84, No. 15, Apr. 1960, pp. 288–295.

Crosby, George, Jr. and Daniel J. Mackay, Jr., "Testing Halon 1301 System Design," *Fire Journal,* Vol. 71, No. 5, Sept., 1977:74 ff.

Darling, Pamela W. and others, "Books in Peril," *Library Journal,* Vol. 101, No. 20, Nov. 15, 1976:2341–2351.

Darling, Pamela W., "Preservation, a National Plan at Last?," *Library Journal,* Vol. 102, No. 4, Feb. 15, 1977:447–449.

Dowling, J. H. and Charles Burton Ford, "Halon 1301 Total Flooding System for Winterthur Museum," *Fire Journal,* Vol. 63, No. 6, Nov. 1969:10–14.

Downs, Barry V., "A Richness of Form, Space and Books," *Canadian Architect,* Jan., 1978:25–29.

Fischer, David J., "Problems Encountered, Hurricane Agnes Flood, June 23, 1972 at Corning, N.Y. and the Corning Museum of Glass," *Conservation Administration.* North Andover, Massachusetts: New England Document Conservation Center, 1975:170–87.

Fischer, David J., Ph.D., and Thomas Duncan, "Conservation Research: Flood-Damaged Library Materials," AIC Bulletin, Vol. 15, No. 2, Summer, 1975:27.

Fischer, Walter R., "Fire Safety Systems: Protecting Our Treasures from Threat of Fire," *Technology and Conservation,* Fall, 1976.

Ford, Charles L., "Halon 1301 Fire-Extinguishing Agent: Properties and Applications," *Fire Journal,* Vol. 64, No. 6, Nov. 1970:36.
Ford, Charles L., "Halon 1301 Update: Research, Application, New Standard," *Specifying Engineer,* May, 1977.
Ford, Charles L., "Winterthur Revisited," NFPA *Fire Journal,* Vol. 69, No. 1, Jan. 1975:81–2.
Goetz, Arthur H., "Books in Peril: A History of Horrid Catastrophes," *Wilson Library Bulletin,* Vol. 47, No. 5, Jan. 1973:428–439.
Hall, Tony, "Letter and Parcel Bomb Precautions," *The Bookseller,* Dec. 7, 1974: 2834.
Harvey, Bruce K., "Fire Hazards in Libraries," *Library Security Newsletter,* Vol. 1, No. 1, Jan. 1975.
Hemphill, B. F., "Lessons of a Fire," *Library Journal,* Vol. 87, No. 6, Mar. 15, 1962:1094–5.
Jensen, Rolf, "Twenty-one Ways to Better Sprinkler System Design," *Fire Journal,* Vol. 68, No. 1, Jan. 1974:47.
Kennedy, J., "Library Arson," *Library Security Newsletter,* Vol. 2, Spring, 1976:1–3.
Lein, H., "Automatic Fire Detection Devices and Their Operating Principles," *Fire Engineering,* June, 1975:38–42.
Levin, Bernard, "Psychological Characteristics of Firesetters," *Fire Journal,* Vol. 70, No. 2, Mar. 1976:36–41.
Lucht, David A., "Basic Considerations: How Safe Is Safe?," *Specifying Engineer,* Vol. 29, No. 5, May, 1978:66–69.
Maddox, Jane, "The Norwood Book Burning," *Wilson Library Bulletin,* Vol. 24, 1960:412–14.
Martin, John H., "Resuscitating a Waterlogged Library," *Wilson Library Bulletin,* Nov. 1975:241–243.
Meisel, George, "The Professional Liability of Architects and Engineers," *Fire Journal,* Vol. 66, No. 4, July, 1972:14.
Merdinyan, Philip H., "A Fully Approved On-Off Sprinkler," *Fire Journal,* Vol. 67, No. 1, Jan. 1973:11–15.
Mittelgluck, E. L., "Case Study in Library Arson," *Library Security Newsletter,* Vol. 2, Spring, 1976.
Morris, John, "Current Fire Protection Topics," *Professional Safety,* Feb. 1975:38–43.
Northey, J., "Halon Extinguishing Agents," *Fire Prevention,* No. 122, Dec. 1977: 22–4.
Patton, Richard, "The Life Safety System," *Fire Journal,* Vol. 65, No. 1, Jan. 1971: 48–52.
Piez, Gladys T., "Insurance and the Protection of Library Resources," ALA Bull., May, 1962:421–424.
Pool, Brady and Roy C. Mabry, "Fire in Unsprinklered Warehouse Destroys Bindery and a Million Books," *Fire Engineering,* Vol. 130, No. 8, Aug. 1976:71–74.
Poole, William F., "The Construction of Library Buildings," *Library Journal,* Vol. 6, No. 4, Apr. 1881:69.
Sager, Don, "Protecting the Library After Hours," *Library Journal,* Vol. 94, No. 18, Oct. 15, 1969:3609–14.
Schaefer, George L., "Fire," *Library Journal,* Vol. 85, No. 3, Feb. 1, 1960:504–5.

Schell, H. B., "Cornell Starts a Fire" *Library Journal,* Vol. 85, No. 17, Oct. 1, 1961:3398.
Schmelzer, Menahem, "Fire and Water: Book Salvage in New York and in Florence," *Special Libraries,* Vol. 59, No. 8, Oct. 1968:620–625.
Sellers, David Y. and Richard Strassberg, "Anatomy of a Library Emergency," *Library Journal,* Vol. 98, No. 17, Oct. 1, 1973:2824–27.
Spawn, Wilman, "After the Water Comes," *P.L.A. Bulletin,* Pennsylvania Library Association, Vol. 28, No. 6, Nov. 1973:243–51.
Stender, Walter and Evans Walker, "The National Personnel Records Center Fire: A Study in Disaster," *American Archivist,* Vol. 37, No. 4, Oct. 1974, 521ff.
Surrency, Erwin C., "Guarding Against Disaster," *Law Library Journal,* Vol. 66, No. 4, Nov. 1973:419–428.
Surrency, Erwin C., "Freeze-Dried Books," *Library Journal,* Vol. 99, No. 16, Sept. 15, 1974:2108–9.
Swayne, Leo H., "Fire Protection at the National Archives Building," NFPA *Fire Journal,* Vol. 69, No. 1, Jan. 1975.
Thompson, Robert J., "The Decision Tree for Fire Safety Systems Analysis: What It Is and How to Use It," *Fire Journal,* July, Sept., Nov. 1975. (Parts I–II–III)
Toffler, Alvin, "Libraries" in *Bricks and Mortarboards,* A Report from Educational Facilities Laboratories, May, 1964:71–98.
Trelles, Oscar M., "Protection of Libraries," *Law Library Journal,* Vol. 66, No. 3, Aug. 1973:241–258.
Tweedie, A. T., "Letters," *Library Journal,* Vol. 99, No. 2, Jan. 15, 1974:82.
Walford, Cornelius, "Destruction of Libraries from Fire Considered Practically and Historically," *Transactions and Proceedings of the Second Annual Meeting of the Library Association of the United Kingdom,* London, 1880:65–70.
Walford, Cornelius, "Chronological Sketch of the Destruction of Libraries by Fire in Ancient and Modern Times, and of Other Severe Losses of Books and Mss. by Fire or Water," Appendix V, *Transactions and Proceedings of the Second Annual Meeting of the Library Association of the United Kingdom,* London, 1880:149–154.
Walker, Evans, "Records Recovery-Salvage of Wet Papers," *Fire Journal,* Vol. 68, No. 4, July, 1974:65–66.
Waters, Peter, "Does Freeze Drying Save Books or Doesn't It? Salvaging a Few Facts from a Flood of (Alleged) Misinformation," *American Libraries,* July–August, 1975:422–3.
Wood, Jack A., "Stop-and-Go Sprinklers," *Fire Journal,* Vol. 61, No. 6, Nov. 1967:86–88.
Yates, Rob, Bonnie Carson, L. E. Jones, Doug Chmara, Jim Burpee, "Engineering Library Survives," and other accounts of the University of Toronto fire, *Toike Oike,* Vol. LXXX, No. 7, Feb. 17, 1977, a special issue of the University of Toronto Undergraduate Engineering Society newspaper.

UNSIGNED ARTICLES:

"The Arson Puzzle: Can the Pieces be Put Together?," *Record,* Vol. 55, No. 2, March–April, 1978.
E.F.L., "Caught in a Book Bind? New Study Reviews Economics of Storage in

Libraries," Educational Facilities Laboratories, *EFL College Newsletter,* June 1969:18–25.
"Fire Rated Top Problem in Security Study," *Library Journal,* Vol. 98, No. 18, Oct. 15, 1973:2954.
"Fire Losses Increase by 205% in Ten Years," *Business Insurance,* June 30, 1975.
"Fires and Fire Traps," *Library Journal,* Vol. 83, No. 10, May 15, 1958:1508.
"Hartford Branch Burned in September Riot," *Library Journal,* Vol. 94, No. 18, 1969:3591.
"Halogenated Extinguishing Agents," *NFPA Quarterly,* Vol. 48-2, Oct. 1954.
"Metropolitan Toronto Library, Architect: Raymond Moriyama," *Canadian Architect,* January, 1978:20–24.
"Modules Bloom in Saudi Desert for $3 Billion University Job," *Engineering News-Record,* Mar. 9, 1978:24–5.
"The Need for Greater Security Against Arson," *Fire Prevention* #125, Fire Protection Association, London EC4N 1TJ, June, 1978:20–23.
"A Needless Disaster," *Public Libraries,* Vol. 16, No. 5, May, 1911:200–1.
"Preservation of Books from Fire," *Library Journal,* Vol. 9, No. 11, Nov. 1884:191.
"The Public Record Office, Kew," *Fire Prevention* #125, Fire Protection Association, London EC4N 1TJ, June, 1978:16–19.
"In Deference to Its Environment, The Pusey Library Was Built Beneath Harvard Yard," *Architectural Record,* Vol. 160, No. 4, Sept. 1976:97–102.
"A System to Monitor a Building's Safety," *Business Week,* Aug. 16, 1976:62B.
"Walker Library Destroyed by Fire . . .," *Cass County Independent,* Walker, MI, April 8, 1976.
"Woman Discovered Dead in Boston Library," *Library Journal,* Vol. 102, No. 18, Oct. 15, 1977:2106.
"Zoned Halon System Overcomes First-Cost Hurdle," *Architectural Record,* Vol. 162, No. 6, Mid-October, 1977:12–13.

TECHNICAL REPORTS

*Austin, Texas, Fire Department, Report of Fire at 3637 Dorsett, Feb. 2–3, 1976.
*Charlottesville, Virginia, Fire Department, Report of Fire at Alderman Library, Aug. 7, 1974.
Chicarello, Peter J., J. M. Troupe and R. K. Dean, "Fire Tests in Mobile Storage Systems for Archival Storage," Technical Report No. J. I. 3A3N4.RR, Norwood, Mass., Factory Mutual Research, June, 1978.
*Harvard Libraries Preservation Committee, to the Harvard University Council, Mar. 16, 1977.
Haskell Laboratory Report, No. 577-74, "Toxicity of Bromotrifluormethane," Haskell Laboratory for Toxicology and Industrial Medicine, Aug. 19, 1974.
Newman, R. Murray, "Record Storage Fire Tests," Norwood, Mass., Factory Mutual Research, Dec. 1974.
*Ontario Fire Marshal Report, Fire of November 12, 1973, St. Clair College of Applied Arts and Technology, Jan. 15, 1974.
*Toronto, Ontario, Fire Department Report, Fire at University of Toronto, February 11, 1977.

*University of Toronto, unpublished chart: "Disaster Contingency Planning: Assignment of Responsibility (for Salvage and Other Operations,") 1976.
*University of Toronto, unpublished chart: "Disaster Contingency Planning: Goods and Services for Salvage Operations," 1976.
*Windsor, Ontario, Fire Department report, Fire at St. Clair College of Arts and Technology, Nov. 13, 1973.

* Unpublished.

The Huntington Library, San Marino, California